Other Books by the Author

Mysteries in Knowledge
Challenge Your Children
Life Lessons
Mental Multiplication Volume 1

MENTAL MULTIPLICATION

VOLUME 2

SHARON BRIGGS

LifeRich Publishing is a registered trademark of The Reader's Digest Association, Inc.

LifeRich Publishing books may be ordered through booksellers or by contacting:

LifeRich Publishing
1663 Liberty Drive
Bloomington, IN 47403
www.liferichpublishing.com
844-686-9607

ISBN: 978-1-4897-4688-7 (sc)
ISBN: 978-1-4897-4687-0 (e)

Print information available on the last page.

LifeRich Publishing rev. date: 06/20/2023

Saturate yourself with His presence and you will never be the same.

The Good Samaritan

Don't be a men pleaser: Jesus was without sin and
was denied and betrayed, and crucified.

You could
Be
Perfect
but let God perfect you. He holds the world in His hands.

He paid for your redemption

Your rescue will never leave behind faith in the living God.

Give and you shall receive

God is your defense

See your victory, receive your victory!

Your mountain can move, with only the faith of a grain of mustard seed.

Do heaven, there is no sin there.

Be you, if your you is in Christ

Dress for the occasion, you're the bride of Christ

You can de-escalate any situation, by remembering who is inside of you.

The life of God is the only thing that works consistently.

Do you have courage to try Jesus? He will never fail you!

Giving God complete glory will make all the difference in the world.

Romans 10:9,10
You can feel like the one who made you.
Receive your Creator into your heart.

Behold I stand at the door and knock.

Overrule and override what should be but what
is not. It is in the heart that counts.

When things get too heavy, hand them over to the
one who has more power than all power!

Trust God

Your rescue is in the Name of the Lord.

A beautifully wrapped package is the one that
you will present to God at His coming.

Come as a little child

To the natural man the spirit will make that man alive.

Your humility will give you the resolve to know you are powerless to help yourself. You must have God.

Your help is in the name of the Lord

Call on Him while He is near

Great is His faithfulness

I went and got my requisition because I had a
receipt: The qualified Word of God.

The manager of the store is Jesus who is the creator of the universe.

Greater is He who is in you than he who is in the world.

See it, believe it, receive it.

Faith comes by hearing, and hearing by the
Word of God. Do you believe the Word of God?

Happy is that people whose God is the Lord

Heaven is not hard to do, there is no sin in that place.

Forgiveness is a golden Key that unlocks the door to God's presence.

Nothing is too hard for God.

The heavier your load the more you are consenting to carry it.

Happy is that people that maketh the Lord their trust

Without Him you can do nothing

You can do all things through Christ who strengthens you.

If you want to see miracles then look deeply into the Word of God.

Do miracles, believe the impossible.
This book: The Bible, is God given, God received.

Heaven is not hard to do. In love and kindness have you been drawn.

CONTENTS

INTRODUCTION

A note to all the readers of Mental Multiplication Vol. 2; it is especially critical that you read and study Mental Multiplication Vol 1. (Barnes and Nobles, Amazon.com) The training in addition and multiplication is essential for your success in Mental Multiplication Vol. 2. Please prayerfully enter into your study of this material.

You will come out of this feeling like a champion. I feel good that the Holy Spirit has taught me so much. Using the calculator was only to confirm my answers, not to achieve my answers. I have the knowledge of a witty invention, and so will you.

PART I

TIME TABLES

1X1=1
1X2=2
1X3=3
1x4=4
1X5=5
1X6=6
1X7=7
1X8=8
1x9=9
1x10=10
1x11=11
1x12=12
2*1=2
2*2=4
2*3=6
2*4=8
2*5=10
2*6=12
2*7=14
2*8=16
2*9=18
2*10=20
2*11=22
2*12=24
3*1=3
3*2=6
3*3=9
3*4=12
3*5=15
3*6=18
3*7=21
3*8=24
3*9=27
3*10=30
3*11=33
3*12=36

Your worse hour can be your greatest victory.
Your tears are preserved in a bottle.

4*1=4
4*2=8
4*3=12
4*4=16
4*5=20
4*6=24
4*7=28
4*8=32
4*9=36
4*10=40
4*11=44
4*12=48
5*1=5
5*2=10
5*3=15
5*4=20
5*5=25
5*6=30
5*7=35
5*8=40
5*9=45
5*10=50
5*11=55
5*12=60
6*1=6
6*2=12
6*3=18
6*4=24
6*5=30
6*6=36
6*7=42
6*8=48
6*9=54
6*10=60
6*11=66
6*12=72

7*1=7
7*2=14
7*3=21
7*4=28
7*5=35
7*6=42
7*7=49
7*8=56
7*9=63
7*10=70
7*11=77
7*12=84
8*1=8
8*2=16
8*3=24
8*4=32
8*5=40
8*6=48
8*7=56
8*8=64
8*9=72
8*10=80
8*11=88
8*12=96
9*1=9
9*2=18
9*3=27
9*4=36
9*5=45
9*6=54
9*7=63
9*8=72
9*9=81
9*10=90
9*11=99
9*12=108

```
10*1=10
10*2=20
10*3=30
10*4=40
10*5=50
10*6=60
10*7=70
10*8=80
10*9=80
10*10=90
10*11=110
10*12=120
11*1=11
11*2=22
11*3=33
11*4=44
11*5=55
11*6=66
11*7=77
11*8=88
11*9=99
11*10=110
11*11=121
11*12=132
12*1=12
12*2=24
12*3=36
12*4=48
12*5=60
12*6=72
12*7=84
12*8=96
12*9=108
12*10=120
12*11=132
12*12=144
```

13*1=13
13*2=26
13*3=39
13*4=52
13*5=65
13*6=78
13*7=91
13*8=104
13*9=117
13*10=130
13*11=143
13*12=156
14*1=14
14*2=28
14*3=42
14*4=56
14*5=70
14*6=84
14*7=98
14*8=112
14*9=126
14*10=140
14*11=154
14*12=168
15*1=15
15*2=30
15*3=45
15*4=60
15*5=75
15*6=90
15*7=105
15*8=120
15*9=135
15*10=150
15*11=165
15*12=180

Who are you? A child of the living God. John 1:12
Come as a little child.

16*1=16
16*2=32
16*3=48
16*4=64
16*5=80
16*6=96
16*7=112
16*8=128
16*9=144
16*10=160
16*11=176
16*12=192
17*1=17
17*2=34
17*3=51
17*4=68
17*5=85
17*6=102
17*7=119
17*8=136
17*9=153
17*10=170
17*11=187
17*12=204
18*1=18
18*2=36
18*3=54
18*4=72
18*5=90
18*6=108
18*7=126
18*8=144
18*9=162
18*10=180
18*11=198
18*12=216.

The fruit of humility is to practice love.

19*1=19
19*2=38
19*3=57
19*4=76
19*5=95
19*6=114
19*7=133
19*8=152
19*9=171
19*10=190
19*11=209
19*12=228
20*1=20
20*2=40
20*3=60
20*4=80
20*5=100
20*6=120
20*7=140
20*8=160
20*9=180
20*10=200
20*11=220
20*12=240
21*1=21
21*2=42
21*3=63
21*4=84
21*5=105
21*6=126
21*7=147
21*8=168
21*9=189
21*10=210
21*11=231
21*12=252

22*1=22
22*2=44
22*3=66
22*4=88
22*5=110
22*6=132
22*7=154
22*8=176
22*9=198
22*10=220
22*11=242
22*12=264
23*1=23
23*2=46
23*3=69
23*4=92
23*5=115
23*6=138
23*7=161
23*8=184
23*9=207
23*10=230
23*11=253
23*12=276
24*1=24
24*2=48
24*3=72
24*4=96
24*5=120
24*6=144
24*7=168
24*8=192
24*9=216
24*10=240
24*11=264
24*12=288

Those who are willing to bend, are also willing to bow.

25*1=25
25*2=50
25*3=75
25*4=100
25*5=125
25*6=150
25*7=175
25*8=200
25*9=225
25*10=250
25*11=275
25*12=300
26*1=26
26*2=52
26*3=78
26*4=104
26*5=130
26*6=156
26*7=182
26*8=208
26*9=234
26*10=260
26*11=286
26*12=312
27*1=27
27*2=54
27*3=81
27*4=1
27*5=135
27*6=162
27*7=189
27*8=216
27*9=243
27*10=270
27*11=297
27*12=324

The finish line is not on earth, but in heaven.

Your mind set is established by your determination to succeed.

28*1=28
28*2=56
28*3=84
28*4=112
28*5=140
28*6=168
28*7=196
28*8=224
28*9=252
28*10=280
28*11=308
28*12=336
29*1=29
29*2=58
29*3=87
29*4=116
29*5=145
29*6=174
29*7=203
29*8=232
29*9=261
29*10=290
29*11=319
29*12=348
30*1=30
30*2=60
30*3=90
30*4=120
30*5=150
30*6=180
30*7=210
30*8=240
30*9=270
30*10=300
30*11=330
30*12=360

Your reconciliation to Christ is called obedience.

31*1=31
31*2=62
31*3=93
31*4=124
31*5=155
31*6=186
31*7=217
31*8=248
31*9=279
31*10=310
31*11=341
31*12=372
32*1=32
32*2=64
32*3=96
32*4=128
32*5=160
32*6=192
32*7=224
32*8=256
32*9=288
32*10=320
32*11=352
32*12=384
33*1=33
33*2=66
33*3=99
33*4=132
33*5=165
33*6=198
33*7=231
33*8=264
33*9=297
33*11=363
33*12=396

Your victory can only be determined by your perseverance.

34*1=34
34*2=68
34*3=102
34*4=136
34*5=170
34*6=204
34*7=238
34*8=272
34*9=306
34*10=340
34*11=374
34*12=408
35*1=35
35*2=70
35*3=105
35*4=140
35*5=175
35*6=210
35*7=245
35*8=280
35*9=315
35*10=350
35*11=385
35*12=420
36*1=36
36*2=72
36*3=108
36*4=144
36*5=180
36*6=216
36*7=252
36*8=288
36*9=324
36*10=360
36*11=396
36*12=432

Grace enables you to be more than a conqueror

37*1=37
37*2=74
37*3=111
37*4=148
37*5=185
37*6=222
37*7=259
37*8=296
37*9=333
37*10=370
37*11=407
37*12=444
38*1=38
38*2=76
38*3=114
38*4=152
38*5=190
38*6=228
38*7=266
38*8=304
38*9=342
38*10=380
38*11=418
38*12=456
39*1=39
39*2=78
39*3=117
39*4=156
39*5=195
39*6=234
39*7=273
39*8=312
39*9=351
39*10=390
39*11=429
39*12=468

40*1=40
40*2=80
40*3=120
40*4=160
40*5=200
40*6=240
40*7=280
40*8=320
40*9=360
40*10=400
40*11=440
40*12=480
41*1=41
41*2=82
41*3=123
41*4=164
41*5=205
41*6=246
41*7=287
41*8=328
41*9=369
41*10=410
41*11=451
41*12=492
42*1=42
42*2=84
42*3=126
42*4=168
42*5=210
42*6=252
42*7=294
42*8=336
42*9=378
42*10=420
42*11=462
42*12=504

43*1=43
43*2=86
43*3=129
43*4=172
43*5=215
43*6=258
43*7=301
43*8=344
43*9=387
43*10=430
43*11=473
43*12=516
44*1=44
44*2=88
44*3=132
44*4=176
44*5=220
44*6=264
44*7=308
44*8=352
44*9=396
44*10=410
44x11=484
44x12=528
45*1=45
45*2=90
45*3=135
45*4=180
45*5=225
45*6=270
45*7=315
45*8=360
45*9=405
45*10=450
45*11=495
45*12=540

46*1=46
46*2=92
46*3=138
46*4=184
46*5=230
46*6=276
46*7=322
46*8=368
46*9=414
46*10=460
46*11=506
46*12=552
47*1=47
47*2=94
47*3=141
47*4=188
47*5=235
47*6=282
47*7=329
47*8=376
47*9=423
47*10=470
47*11=517
47*12=564
48*1=48
48*2=96
48*3=144
48*4=192
48*5=240
48*6=288
48*7=336
48*8=384
48*9=432
48*10=480
48*11=528
48*12=576

49*1=49
49*2=98
49*3=147
49*4=196
49*5=245
49*6=294
49*7=343
49*8=392
49*9=441
49*10=490
49*11=539
49*12=588
50*1=50
50*2=100
50*3=150
50*4=200
50*5=250
50*6=300
50*7=350
50*8=400
50*9=450
50*10=500
50*11=550
50*12=600
51*1=51
51*2=102
51*3=153
51*4=204
51*5=255
51*6=306
51*7=357
51*8=408
51*9=459
51*10=510
51*11=561
51*12=612

52*1=52
52*2=104
52*3=156
52*4=208
52*5=260
52*6=312
52*7=364
52*8=416
52*9=468
52*10=520
52*11=572
52*12=624
53*1=53
53*2=106
53*3=159
53*4=212
53*5=265
53*6=318
53*7=371
53*8=424
53*9=477
53*10=530
53*11=583
53*12=636
54*1=54
54*2=108
54*3=162
54*4=216
54*5=270
54*6=324
54*7=378
54*8=432
54*9=486
54*10=540
54*11=594
54*12=648

55*1=55
55*2=110
55*3=165
55*4=220
55*5=275
55*6=330
55*7=385
55*8=440
55*9=495
55*10=550
55*11=605
55*12=660
56*1=56
56*2=112
56*3=168
56*4=224
56*5=280
56*6=336
56*7=392
56*8=448
56*9=504
56*10=560
56*11=616
56*12=672
57*1=57
57*2=114
57*3=171
57*4=228
57*5=285
57*6=342
57*7=399
57*8=456
57*9=513
57*10=570
57*11=627
57*12=684

58*1=58
58*2=116
58*3=174
58*4=232
58*5=290
58*6=348
58*7=406
58*8=464
58*9=522
58*10=580
58*11=638
58*12=696

59*1=59
59*2=118
59*3=177
59*4=236
59*5=295
59*6=354
59*7=413
59*8=472
59*9=531
59*10=590
59*11=649
59*12=708

60*1=60
60*2=120
60*3=180
60*4=240
60*5=300
60*6=360
60*7=420
60*8=480
60*9=540
60*10=600
60*11=660
60*12=720
61*1=61
61*2=122
61*3=183
61*4=244
61*5=305
61*6=366
61*7=427
61*8=488
61*9=549
61*10=610
61*11=671
61*12=732
62*1=61
62*2=124
62*3=186
62*4=248
62*5=310
62*6=372
62*7=434
62*8=496
62*9=558
62*10=620
62*11=682
62*12=744

63*1=63
63*2=126
63*3=189
63*4=252
63*5=315
63*6=378
63*7=441
63*8=504
63*9=567
63*10=630
63*11=693
63*12=756
64*1=64
64*2=128
64*3=192
64*4=256
64*5=320
64*6=384
64*7=448
64*8=512
64*9=576
64*10=640
64*11=704
64*12=768
65*1=65
65*2=130
65*3=195
65*4=260
65*5=325
65*6=390
65*7=455
65*8=520
65*9=585
65*10=650
65*11=715
65*12=780

66*1=66
66*2=132
66*3=198
66*4=264
66*5=330
66*6=396
66*7=462
66*8=528
66*9=594
66*10=660
66*11=726
66*12=792
67*1=67
67*2=134
67*3=201
67*4=268
67*5=335
67*6=402
67*7=469
67*8=536
67*9=603
67*10=670
67*11=737
67*12=804
68*1=68
68*2=136
68*3=204
68*4=272
68*5=340
68*6=408
68*7=476
68*8=544
68*9=612
68*10=680
68*11=748
68*12=816

```
69*1=69
69*2=138
69*3=207
69*4=276
69*5=345
69*6=414
69*7=483
69*8=552
69*9=621
69*10=690
69*11=759
69*12=828
70*1=70
70*2=140
70*3=210
70*4=280
70*5=350
70*6=420
70*7=490
70*8=560
70*9=630
70*10=700
70*11=770
70*12=840
71*1=71
71*2=142
71*3=213
71*4=284
71*5=355
71*6=426
71*7=497
71*8=568
71*9=639
71*10=710
71*11=781
71*12=852
```

72*1=72
72*2=144
72*3=216
72*4=288
72*5=360
72*6=432
72*7=504
72*8=576
72*9=648
72*10=720
72*11=792
72*12=864
73*1=73
73*2=146
73*3=219
73*4=292
73*5=365
73*6=438
73*7=511
73*8=584
73*9=657
73*10=730
73*11=803
73*12=876
74*1=74
74*2=148
74*3=222
74*4=296
74*5=370
74*6=444
74*7=518
74*8=592
74*9=666
74*10=740
74*11=814
74*12=888

75*1=75
75*2=150
75*3=225
75*4=300
75*5=375
75*6=450
75*7=525
75*8=600
75*9=675
75*10=750
75*11=825
75*12=900
76*1=76
76*2=152
76*3=228
76*4=304
76*5=380
76*6=456
76*7=532
76*8=608
76*9=684
76*10=760
76*11=836
76*12=912
77*1=77
77*2=154
77*3=231
77*4=308
77*5=385
77*6=462
77*7=539
77*8=616
77*9=693
77*10=770
77*11=847
77*12=924

78*1=78
78*2=156
78*3=234
78*4=312
78*5=390
78*6=468
78*7=546
78*8=624
78*9=702
78*10=780
78*11=858
78*12=936
79*1=79
79*2=158
79*3=237
79*4=316
79*5=395
79*6=474
79*7=553
79*8=632
79*9=711
79*10=790
79*11=869
79*12=948
80*1=80
80*2=160
80*3=240
80*4=320
80*5=400
80*6=480
80*7=560
80*8=640
80*9=720
80*10=800-
80*11=880
80*12=960

81*1=81
81*2=162
81*3=243
81*4=324
81*5=405
81*6=486
81*7=567
81*8=648
81*9=729
81*10=810
81*11=891
81*12=972
82*1=82
82*2=164
82*3=246
82*4=328
82*5=410
82*6=492
82*7=574
82*8=656
82*9=738
82*10=820
82*11=902
82*12=984
83*1=83
83*2=166
83*3=249
83*4=332
83*5=415
83*6=498
83*7=581
83*8=664
83*9=747
83*10=830
83*11=913
83*12=996

84*1=84
84*2=168
84*3=252
84*4=336
84*5=420
84*6=504
84*7=588
84*8=672
84*9=756
84*10=840
84*11=924
84*12=1,008
85*1=85
85*2=170
85*3=255
85*4=340
85*5=425
85*6=510
85*7=595
85*8=680
85*9=765
85*10=850
85*11=935
85*12=1,020
86*1=86
86*2=172
86*3=258
86*4=344
86*5=430
86*6=516
86*8=688
86*9=774
86*10=860
86*11=946
86*12=1,032

87*1=87
87*2=174
87*3=261
87*4=348
87*5=435
87*6=522
87*7=609
87*8=696
87*9=783
87*10=870
87*11=957
87*12=1,044
88*1=88
88*2=176
88*3=264
88*4=352
88*5=440
88*6=528
88*7=616
88*8=704
88*9=792
88*10=880
88*11=968
88*12=1,056
89*1=89
89*2=178
89*3=267
89*4=356
89*5=445
89*6=534
89*7=623
89*8=712
89*9=801
89*10=890
89*11=979
89*12=1,068

```
90*1=90
90*2=180
90*3=270
90*4=360
90*5=450
90*6=540
90*7=630
90*8=720
90*9=810
90*10=900
90*11=990
90*12=1,080
91*1=91
91*2=182
91*3=273
91*4=364
91*5=455
91*6=546
91*7=637
91*8=728
91*9=819
91*10=910
91*11=1,001
91*12=1,092
92*1=92
92*2=184
92*3=276
92*4=368
92*5=460
92*6=552
92*7=644
92*8=736
92*9=828
92*10=920
92*11=1,012
92*12=1,104
```

93*1=93
93*2=186
93*3=279
93*4=372
93*5=465
93*6=558
93*7=651
93*8=744
93*9=837
93*10=930
93*11=1,023
93*12=1,116
94*1=94
94*2=188
94*3=282
94*4=376
94*5=470
94*6=569
94*7=658
94*8=752
94*9=846
94*10=940
94*11=1,034
94*12=1,128
95*1=95
95*2=190
95*3=285
95*4=380
95*5=475
95*6=570
95*7=665
95*8=760
95*9=855
95*10=950
95*11=1,045
95*12=1,140

96*1=96
96*2=192
96*3=288
96*4=384
96*5=480
96*6=576
96*7=672
96*8=768
96*9=864
96*10=960
96*11=1,056
96*12=1,152
97*1=97
97*2=194
97*3=291
97*4=388
97*5=485
97*6=582
97*7=679
97*8=776
97*9=873
97*10=970
97*11=1,067
97*12=1,164
98*1=98
98*2=196
98*3=294
98*4=392
98*5=490
98*6=588
98*7=686
98*8=784
98*9=882
98*10=980
98*11=1,078
98*12=1,176

99*1=99
99*2=198
99*3=297
99*4=396
99*5=495
99*6=594
99*7=693
99*8=792
99*9=891
99*10=990
99*11=1,089
99*12=1,188

100

PART II

EXPLANATIONS

Mental Multiplication
Time Tables Explanations: 1-9

1. 1*1-1-9 Single digit explanations
   1*1=1,1*2=2,1*3-1*9: Self Explanatory

2. 2*1=2, 2*2=4=8 2*3=6-2*9=18 Explanation: 2*8=16 (8+8=16. Finger counting by 2"s 2,4,6,8,10,12(fingers in order, then use thumbs for "11" and "12").

3. 3*1=3 finger count by "3's.: 3,6,9,12,15,18,21,24,27,30,33,36.

4. 4*1=4 finger count by "4's" 4,8,12,16,20,24,28,32,36,40,44,48

5. 5*1=5 finger count by "5's" 5,10, 15,20,20,25 30, 35,40.45,50,55,60.

   cont.

6. 6*1=6 finger count by "6's 6,12,18,24 30,36,42,48,54,60,66,72

7. 7*1=7 finger count by "7"s 7,14,21,28,35,42,49,56,63,70,77,84.

8. 8*1=8 finger count by "8's" 8, 16,24,32,40, 48, 56, 64,72,80,88,96.

9. 9*1=9 finger count by "9's" 9,18, 27,36,45,54,63,72,81,90,99,108.

Beginning of 2 digit time tables.

1. 10*1=10 simply add "0" to the time table answer i.e. 10*2=20
   10*3=30, 10*4=40, 10*5=50, 10*6=60, 10*7=70, 10*8=80,10*9=90, 10*10=100.

2. 11 time tables 1-9 (double the digits in the final answer) i.e. 11*1=11- 11*2=22 11x3=33
   11*4=44 11*5=55, 11*6=66, 11*7=77, 11*8=88, 11*9=99, 11*10=110,11*11=121,11*12=132

3. More examples: "11" multiplied, double the numbers, i.e. 11*1=11, 11*2=22, 11*3=33 etc. But after "9" i.e. 10 *11= 110, (expanded) 1*0=0, 1*1 (diagonally)=1, 1*1 vertically)=1
   Final Answer: 110
    10
   X11
   110

4.

```
   11
  X11
  121
```
Pattern: 1. 1*1=1 (vertical left)

2. 1*1+1=2 (diagonal right)

3. 1*1=1 (vertical left)) Final Answer: 121

(Use this pattern for double digit multiplication)

```
   11
  x12
  132
```
Your turn

```
   12
  X 4
   48
```
12's Explanations

12*1=12, 12*2=24 12*3=36

Pattern: 12*4= 4*2=8,4*1=4 Final answer: "48"

13 Explanations. 13*1=13, 13*2=26, 13*3=39
Pattern: 13*4=52. 4*3=12, put down 2, carry 1. 4*1+1=5 Final answer: 52

```
   13
  X4
  52
```

14 Explanations 14*1=14, 14*2=28,14*3=52, 14*4=56
Pattern: 14*4=56 4*4=16, add the "1" to 4=5 Final Answer: 56

```
   14
  X4
  56
```

Explanations: 15*1=15, 15*2=30, 15*3=45 15*4=60
Pattern: 15*4=60, 4*5=20, put down "0" add the "2" +4*1=6 Final answer:60 i.e. x 4
Pattern: 20*4=80 4*0=0, 4*2=8 Final answer= 80 21x1=21, 21x2=42 21x3= 63, 21x4=84.
21x5=105 21x6=126, 21x7=147, 21x8=168, 21x9=189, 21x10=210, 21x11=231, 21x12=252.
22x1=22, 22x2=24,22x5=110,22x6=132, 22x7=154, 22x8=176,22x9=198, 22x10=220,
22x11=22x11=242, 22x12,=264. 23*1=23, 23*2=46, 23*3=69, etc.
Pattern: 23*4=92, 3*4=12 put down the "2" add the "1" to the 4*2 =9. Final answer: 92

5. 25*1=25, 25*2=50 25*3=75, 25*4=100
   Pattern: 25*5= 125, 5*5=25, put down "2" add 5 to 12*2=10+2=12
   Final answer: 125. 63*12=756,

   25
   x5
   125

   63*13=819 Pattern: 3*3=9, 3*6+3=21 put down "1" add 2 to "6"=8 Final answer: 819

6. 83*4=332 Pattern: 4*3=12 put down "2," 4x8+1=33 Final Answer: 332

7. 93*7=651 Pattern: 7*3=21, put down "1" Multiply 7*9=63 add 2= Final answer: 651

Please review each grouping of numbers 1-4 until viewer is comfortable with how to obtain the "Final answers."

Your words and your thoughts walk hand in hand.

PART III

SINGLE DIGIT
MULTIPLICATION

1- to 9 digit computations of multiplication. Full Patterns and Steps for two-digit multiplication problems will be in Part 4 of this manual

Note: 1x1-1x9 is self-explanatory 2-9 single digit multiplication

1.	2.	3.	4.	5.	6.	7.	8.	9.
2	2	2	2	2	2	2	2	2
x 1	x2	x3	x4	x5	x6	x7	x8	x9
2	4	6	8	10	12	?	?	?

10	11	12	13	14	15	16	17	18
x2	x 2	x2	x2	x2	x2	x2	x2	x2
20	22	?	26	?	?	32	?	36

Practice: Put the questions with the answers: 14,16,32,34, 28, 18

Sincerely call on the Lord in your time of trouble. He will answer you.

20 21 22   23 24 25 26 27   28 29
X2 x2 x2   x2 x2 x2 x2 x2   x2 x2   Practice: Put questions with answers: 46, 54, 56
40 ?   44   ? 48 50 52 ?   ?   58

30 31 32 33 34 35 36 37 38 39 40 41 42 43 44 45
x2 x2 x2 x2 x2 x2 x2 x2 x2 x2 x2 x2 x2 x2 x2 x2
60 62 64 66 ? 70 ? ?   76 78 80 ?   ?   ?   88 90

Practice: match answers to questions 68,86,82, 84,72,74 (place answers under the correct problems)

42	43	44	45	46	47	48	49
x2	x3	x 4	x5	x6	x7	x8	x9
84	129	176	225	276	329	384	441

53	53	53	53	53	53	53	53	53
X1	x2	x3	x4	x5	x6	x7	x8	x9
53	106	159	212	265	318	371	424	477

54	54	54	54	54	54	54	54	54
x1	x2	x3	x4	x5	x6	7x	8x	x9
54	108	162	216	270	324	378	432	486

60	60	60	60	60	60	60	60	60
X1	x2	x3	x4	x5	x6	x7	x 8	x9
60	120	180	240	300	360	420	480	540

7, 1-9

7	7	7	7	7	7	7	7	7
x1	x2	x3	x4	x5	6x	x7	x8	x9
7	14	21	28	35	42	49	56	63

```
  8    8      8       8      8   8     8   8    8
 x1   x2     x3      x4     x5  6x    x7  x8   x9
  8   16     24      32     40  48    56  64   72


  9  9  9      9      9     9     9  9        9
 x1 x2 x3     x4     x5    6x    x7 x8       x9
  9 18 27     36     45    54    63 72       81
```

10- self-explanatory (just add "0" to the numeric answer.)

PART IV

2 DIGIT MULTIPLICATION STARTS

11-35

1-9 "11" Self-Explanatory, 1-9 add duplicate number.

11	11	11	11	11	11	11	11	11
x1	x2	x3	x4	x5	x6	x7	x8	x9
11	22	33	44	55	66	77	88	99

11	11	11	11	11	11	11	11	11	11
x10	x 11	x 12	x 13	x 14	x 15	x16	x17	x18	x19
110	121	132	143	154	165	176	187	198	209

11	11	11	11	11	11	11	11	11	11
20	x21	x22	x23	x24	x25	x26	x27	x28	x29
220	231	242	253	264	275	286	297	308	319

11	11	11	11	11	11
x30	x31	x32	x33	x34	x35
330	341	352	363	374	385

12	12	12	12	12	12	12	12	12	12
x1	x 2	x3	x4	x5	x6	x7	x8	x9	x10
12	24	36	48	60	72	84	96	108	120

12	12	12	12	12	12	12	12	12
11	12	13	14	15	16	17	18	19
132	144	156	168	180	192	204	216	228

12	12	12	12	12	12	12	12	12
20	21	22	23	24	25	26	27	28
240	252	264	276	288	300	312	324	336

12	12	12	12	12	12	12
29	30	31	32	33	34	35
348	360	372	384	396	408	420

Explanation of single-digit multiplication:

```
 12      1.  2x9=18 (Put down 8, carry 1)
 X9      2.  9x1+1=10
108      3.  Final answer: 108.
```

```
 27      1.  4x7=28 (Put down your 8, carry the 2)
 X4      2.  4x2+2=10 (Put down 10)
108      3.  Final answer: 108
```

12	12	12	12	12	12	12	12	12	12
x10	x11	x12	x13	x 14	x15	x16	x17	x18	x19
120	132	144	156	168	180	192	204	216	228

12	12	12	12	12	12	12	12	12	12
x20	x21	x22	x 23	x 24	x25	x26	x27	x28	x29
240	252	264	276	288	300	312	324	336	348

12	12	12	12	12	12
x30	x 31	x 32	x 33	x 34	x 35
360	372	384	396	408	42

13-35

13	13	13	13	13	13	13	13	13
x1	x 2	x 3	x 4	x 5	x 6	x 7	x 8	x 9
13	26	39	52	65	78	91	104	117

13	13	13	13	13	13	13	13	13	13
x10	x11	x 12	x13	x 14	x15	x16	x17	x18	x19
130	143	156	169	182	195	208	221	234	247

Explanation No. 1

	13	1. 9x3=2<u>7</u>
	X19	2. 9x1+2=11+1 x 3=1<u>4</u>
		3. 1x1+1=<u>2</u>
		Answer: 247

13	13	13	13	13	13	13	13	13	13
x20	x21	x22	x23	x24	x25	x26	x27	x28	x29
260	273	286	299	312	325	338	351	264	377

13	13	13	13	13	13
x30	x31	x 32	x 33	x34	x 35
390	403	416	429	442	455

14-35

14	14	14	14	14	14	14	14	14
X1	x2	3x	x4	x5	x6	x7	x8	x9
14	28	42	56	70	84	98	112	126

14	14	14	14	14	14	14	14	14	14
X10	x11	x12	x13	x14	x15	x16	x17	x18	x19
140	154	168	182	196	210	224	238	252	266

Explanation:	14	4x9=3<u>6</u>
	x 19	9x1+3+4=1<u>6</u>
		1x1+1=<u>2</u>
		Answer: 266

14	14	14	14	14	14	14	14	14	14
x20	x21	x22	x23	x24	x25	x26	x27	x28	x29
280	294	308	322	336	350	364	378	392	406

14	14	14	14	14
x 30	x 31	x 32	x 33	x35
420	434	448	462	490

15 to 35

15	15	15	15	15	15	15	15	15
x1	x2	x3	x4	x5	x6	x7	x8	x9
15	30	45	60	75	90	105	120	135

15	15	15	15	15	15	15	15	15	15
x10	x11	x12	x13	x 14	x15	x16	x 17	x 18	x 19
150	165	180	195	210	225	240	255	270	285

15	15	15	15	15	15	15	15	15	15
x20	x21	x 22	x23	x24	x25	x26	x27	x28	x29
300	315	330	345	360	375	390	405	420	435

15	15	15	15	15	15
x30	x31	x32	x33	x34	x 35
450	465	480	495	510	525

Explanation:

15	1. 9x5=4<u>5</u>
x19	2. 9x1+4+1 x 5=1<u>8</u>
285	3. 1x1+1=<u>2</u>

| 57 |

16-35

16	16	16		16	16	16	16	16	16
x1	x2	x3		x4	x5	x6	x7	x8	x9
16	32	48		64	80	96	112	128	144

16	16	16		16	16	16	16	16	16	16
X10	x11	x12		x13	x14	x15	x16	x17	x18	x19
160	176	192		208	224	240	256	272	288	304

Explanation

$$\begin{array}{r} 16 \\ \times\ 19 \\ \hline 304 \end{array}$$

1. 9x6=5<u>4</u>
2. 9x1+5+6=2<u>0</u>
3. 1x1+2=<u>3</u>

16	16	16	16	16	16	16	16	16	16
x20	x21	x22	x23	x24	x25	x26	x27	x28	x29
320	336	352	368	384	400	416	432	448	464

16	16	16	16	16	16
x30	x31	x32	x33	x34	x35
480	496	512	528	544	560

17-35

17	17	17	17	17	17	17	17	17
X1	x2	x3	x4	x5	x6	x7	x8	x9
17	34	51	68	85	102	119	136	153

17	17	17	17	17	17	17	17	17	17
x10	x11	x 12	x 13	x14	x15	x16	x17	x18	x19
170	187	204	221	238	255	272	289	306	323

Explanation

$$\begin{array}{r} 17 \\ \underline{\times 19} \\ 323 \end{array}$$

1. 9x7=6<u>3</u>
2. 9x1+6+7=2<u>2</u>
3. 1x1+2=<u>3</u>

18-35

18	18	18	18	18	18	18	18	18
x1	x2	x3	x4	x5	x6	x7	x8	x9
18	36	54	72	90	108	126	144	162

18	18	18	18	18	18	18	18	18	18
x10	x11	x12	x13	x14	x15	x16	x17	x18	x19
180	198	216	234	252	270	288	306	324	342

Explanation:

$$\begin{array}{r} 18 \\ \underline{\times 19} \end{array}$$

1. 9x8=7<u>2</u>
2. 9x1+7+1x8=2<u>4</u>
3. 1x1+2=<u>3</u>

Answer: 342

18	18	18	18	18	18	18	18	18	18
x20	x21	x22	x23	x24	x25	x26	x27	x28	x29
360	378	396	414	432	450	468	486	504	522

18	18	18	18	18	18
x30	x31	x32	x33	x34	x35
540	558	576	594	612	630

19-35

1	2	3	4	5	6	7	8	9

19	19	19	19	19	19	19	19	19
x1	x2	x3	x4	x5	x6	x7	x8	x9
19	38	57	76	95	114	133	152	171

19	19	19	19	19	19	19	19	19	19
X10	x11	x12	x13	x14	x15	x16	x17	x18	x19
190	209	228	247	266	285	304	323	342	361

19	19	19	19	19	19	19	19	19	19
x20	x21	x22	x23	x24	x25	x26	x27	x28	x29
380	399	418	437	456	475	494	513	532	551

19	19	19	19	19	19
x30	x31	x32	x33	x34	x35
570	589	608	627	646	665

Explanation: 19   1. 9x9=8<u>1</u>

              X<u>29</u>   2. 9x1+8=17+2x9=3<u>5</u>

                      3. 2x1+3=<u>5</u>

                    Answer: 551

20-35

20	20	20	20	20	20	20	20	20
x1	x2	x3	x4	x5	x6	x7	x8	x9
20	40	60	80	100	120	140	160	180

20	20	20	20	20	20	20	20	20	20
x10	x11	x12	x13	x14	x15	x16	x17	x18	x19
200	220	240	260	280	300	320	340	360	380

Notice: 10x2=20, double 10,11x2=22 double 11. 12x2=24 double 12 etc.

24	25	26	27	28	29
x20	x20	x 20	x20	x20	x20
480	500	520	540	560	580

20	20	20	20	20	20
x30	x31	x32	x33	x34	x35
600	620	640	660	680	700

21-25   22-25

21	21	21	21	21		22	22	22	22	441
X21	x22	x23	x24	x25		x21	x22	x23	x24	x22
441	462	483	504	525		462	484	506	528	9702

441
22
9702

1. 2x1=2
2. 2x4+2=10=+1
3. 1+2x4+4x2=17
4. 1+4x2=9 Final Answer: 9702

23-35

23	23	23	23	23	23	23	23	23
x1	x2	x3	x4	x5	x6	x7	x8	x9
23	46	69	92	115	138	161	184	207

23	23	23	23	23	23	23	23	23
x10	x11	x12	x13	x14	x15	x16	17	18
230	253	276	299	322	345	338	391	414

23	23	23	23	23	23	23	23	23	23
x20	x21	x22	x23	x24	x25	x26	x27	x28	x29
460	483	506	529	552	575	598	621	644	667

23	23	23	23	23	23	Explanation:	23	1. 9x3=2<u>7</u>
x30	x31	x32	x33	x34	x35		x29	2. 9x2+2x3+2=2<u>6</u>
690	713	736	759	782	805			3. 2x2+2=<u>6</u>
								Answer: 667

24: 1-12

24	24	24	24	24	24	24	24	24	24	24
x1	x 2	x3	x 4	x 5	x6	x7	x8	x10	x 11	x 12
24	48	72	96	120	144	168	192	240	264	288

25: 1-12

25	25	25	25	25	25	25	25	25	25	25	25	Note: Count by 25's:
x1	x2	x3	x4	x5	x6	x7	x8	x9	x10	x11	x12	25, 50, 75, 100,
25	50	75	100	125	150	175	200	225	250	275	300	

26: 1-12

26	26	26	26	26	26	26	26	26	26	26	26
X1	x2	x3	x4	x5	x6	x7	x8	x9	x10	x11	x12
26	52	78	104	130	156	182	208	234	260	286	312

27: 1-9

27	27	27	27	27	27	27	27	27	Explanation: 27	1. 9x7=6<u>3</u>
X1	x2	x3	x4	x5	x6	x7	x8	x9	x9	2. 9x2+6=<u>24</u>
27	54	81	108	135	162	189	216	243	Answer:243	

28: 10-19

28	28	28	28	28	28	28	28	28	28
x10	x11	x12	x13	x14	x15	x16	x17	x18	x19
280	308	336	364	392	420	448	476	504	532

Explanation of single/double digit multiplication:

27   1. 4x 7=28 (Put down your 8, carry the 2)
X4   2. 4x2+2=10 (Put down 10)
108 3. Final answer: 108

Single/Double Digit Explanation: 30's

30   1. 8x0=0
X8   2. 8x3=24
240 3. Final Answer: 240

Explanations:

30   1.9x0=0   31 1. 0x1=0   32   1.1x2=2
X9   2.9x3=27   x10   2.1x1=1   x11   2.1x2+1 x 3=5
270 3.Final Answer: 270   310 3.1x3=3   352   3.1x3=3
                      Final Answer: 310   Final Answer: 352

29: 20-35

29	29	29	29	29	29	29	29	29	29
x20	x21	x22	x23	x24	x25	x26	x27	x28	x29
580	609	638	667	696	725	754	783	812	841

29	29	29	29	29	29
x30	x 31	x 32	x33	x 34	x35
870	899	928	957	986	1,015

30: 1-9

30	30	30	30	30	30	30	30	30	Note: add "0" to what you know
x1	x2	x3	x4	x5	x6	x7	x8	x9	
30	60	90	120	150	180	210	240	270	

31: 10-19

31	31	31	31	31	31	31	31	31	31
x10	x11	x12	x13	x14	x15	x16	x17	x18	x19
310	341	372	403	434	465	496	527	558	589

2-Digit Multiplication Explanation 31: (look at the ending number)

```
  31    1.  8x1=8
 X18    2.  8x3=24+1=25 (Put down 5 carry 2)
 558    3.  1x3+2=5
        4.  Final Answer: 558
```

```
  31    1.  9x1=9
 X19    2.  9x3=27+1=28 (Put down 8 carry 2)
 589    3.  1x3+2=5 (put down 5)
        4.  Final Answer: 589
```

32: 20-32

32	32	32	32	32	32	32	32	32	32
x20	x21	x22	x23	x24	x25	x26	x27	x28	x29
640	672	704	736	768	800	832	864	896	928

33: 30-35

33	33	33	33	33	33
x30	x31	x32	x33	x34	x35
990	1,023	1,056	1,089	1,122	1,155

34: 1-9

34	34	34	34	34	34	34	34	34
x1	x2	x3	x4	x5	x6	x7	x8	x9
34	68	102	136	170	204	238	272	306

35	35	35	35	35	35	35	35	35
x1	x2	x3	x4	x5	x6	x7	x8	x9
35	70	105	140	175	210	245	280	315

35	35	35	35	35	35	35	35	35	35
x10	x11	x12	x13	x14	x15	x16	x17	x18	x19
350	385	420	455	490	525	560	595	630	665

35	35	35	35	35	35	35	35	35	35
x20	x21	x22	x23	x24	x25	x26	x27	x28	x29
700	735	770	805	840	875	910	945	980	1,015

Explanation: two digit cont.

35   1. 9x5=4<u>5</u>
<u>X19</u> 2. 9x3+4=31+1x5=3<u>6</u>
     3. 1x3+3= Answer: 665

35: 30-35

35	35	35	35	35	35
x30	x31	x32	x33	x34	x35
1,050	1,085	1,120	1,155	1,190	1,225

<u>36: 1-12</u>

36	36	36	36	36	36	36	36	36	36	36	36
x1	x2	x3	x4	x5	x6	x7	x8	x9	x10	x11	x12
36	72	108	144	180	216	252	288	324	360	396	432

37: 1-12

37	37	37	37	37	37	37	37	37	37	37	37
x1	x2	x3	x4	x5	x6	x7	x 8	x 9	x10	x11	x12
37	74	111	148	185	222	259	296	333	370	407	444

38: 1-12

38	38	38	38	38	38	38	38	38	38	38	38 (Try Mastery:)
X1	x2	x3	x4	x5	x6	x7	x8	x9	x10	x11	x12
38	76	114	152	190	228	266	304	342	380	418	456

39	39	39	39	39	39	39
X10	x11	x 12	x13	x14	x15	16
390	429	468	507	546	585	624

Explanation of two-digit multiplication cont.

Questions 11: 10-12

1.   11   1. "0" x "1"=0
      x10   2. "1" x 1=1 (Put down 1)
      110   3. 1x1=1 (Put down 1)
              4. Final answer: 110

2.   11   1. "1" x1=1
      x11   2. "1" x1+ "1"x1=2
      121   3. "1"x1=1
              4. Final answer: 121 (1x1=1 is always definite. Notice criss-cross pattern "X.")

3.   11   1. "1" x 2=2
      x12   2. "2x1+1x1=3
      132   3. "1
              4. Final answer: 132

Questions 12: 10-12

    10   1. "2"x0=0
    X12   2. "2"x1=2
    120   3. "1"x1=1
              4. Final answer: 120

    11   1. 3x1=3
    x 13   2. 3x1+1x1=4
    143   3. 1x1=1
              4. Final answer: 143 (remember the criss-cross pattern "X")

```
 12    1.  "2"x2=4
x12    2.  "2x1+1x2=4
 144   3.  "1"1x1=1
       4.  Final answer: 144
```

Questions 15: 10-12

```
 15    1.  0x5=0
x10    2.  1x5=5
150    3.  1x1=1
       4.  Final answer: 150
```

```
 15    1.  5x1=5
X11    2.  1x1+1 x 5=6
165    3.  1x1=1
       4.  Final answer: 165
```

```
 15    1.  2x5=10 (put down "0" carry "1")="0" *
X12    2.  2x1+1+5=8
180    3.  1x1=1
       4.  Final answer: 180
```

PART V

3 DIGIT
MULTIPLICATION

Beginning of 3-digit Multiplication

1.	2.	3.	4.	5.	6.	7.	8.	9.
111	111	111	111	111	111	111	111	111
x11	x12	x13	x14	x15	x16	x17	x18	x19
1,221	1,332	1,443	1,554	1,665	1,776	1,887	1,998	2,109

1	2	3	4	5	6	7	8	9	10
111	111	111	111	111	111	111	111	111	111
X20	x21	x22	x23	x24	x25	x26	x27	x28	x29
2,220	2,331	2,442	2,553	2,664	2,775	2,886	2,997	3,108	3,219

111	111	111	111	111	111
X30	x31	x32	x33	x34	x35
3,330	3,441	3,552	3,663	3,774	3,885

Explanation 3 and 2 digit Multiplication:

$$111$$
$$\underline{\times 11}$$
$$1,221$$

1. $1 \times 1 = 1$
2. $1 \times 1 + 1 = 2$
3. $1 \times 1 + 1 = 2$
4. $1 \times 1 = 1$

Final Answer: 1,221

$$121$$
$$\underline{\times 11}$$
$$1331$$

1. $1 \times 1 = 1$
2. $1 \times 2 + 1 = 3$
3. $1 \times 1 + 2 = 3$
4. $1 =$ Final Answer: 1331

3-digit Multiplication cont.

```
999 Example:    1.  9x9=81
  X999          2.  9x9+ 8+9x9=170
 998001         3.  81+17=98+81=179+81=260
                4.  81+26=107+81=188
                5.  81+18=99
```

Final Answer: 998,001

3-digit multiplication count.

111	121	131	141	151	161	3-digit Multiplication
x111	x111	x111	x111	x111	x111	
12321	13431	14541	15651	16761	17871	

333	444		555	666	777	888
x 333	x 444		x555	x666	x777	x 888
110,889	197,136		308,025	443,556	603,729	788,544


```
999 Explanation    1.  9x9=81
   X999            2.  9x9+8+9x9=170
  998,001          3.  81+17=98+81=179+81=260
                   4.  81+26=107+81=188
                   5.  81+18 = 99=Final Answer: 998,001
```


Examples cont.

222	222	222	222	222	222	222	222	222
x111	x222	x333	x444	x555	x666	x777	x888	x999
24,642	49,284	73,926	98,568	123,210	147,852	172,494	197,136	221,778

3-digit multiplication explanation:

```
  161        1.  1x1=1
X111         2.  1x6+1x1=7
17,871       3.  1x1+1x1+1x6=8
             4.  1x1+1x6=7
             5.  1x1=1 Final Answer: 17,871
```

3 digit multiplication

```
   999
x 999
 998,001
```

3 digit Explanation

```
  222        1.  2x9=18 (Put down 8 carry 1)=8
X999         2.  2x9+2x9+1=37 (Put down 7 carry 3)=7
221,778      3.  2x9+2 x 9+2 x 9+3=57 (Put down 7 carry 5)=7
             4.  2x9+2x9+5=41 (Put down1 carry 4)=1
             5.  2x9+4=22
             6.  Final Answer: 221,778
```

3-digit begins:

333	333	333	333	333	333	333	333	333	333
x100	x 111	x122	x133	x144	x155	x166	x177	x188	x199
33,300	36,963	40,626	44,289	47,952	51,615	55,278	58,941	62,604	66,267

444	444	444	444	444	444	444	444	444	444
x200	x211	x221	x231	x241	x251	x261	x271	x281	x291
88,800	93,684	98,124	102,564	107,004	111,444	115,884	120,324	124,764	129,204

PART VI

4 DIGIT
MULTIPLICATION

4-digit Multiplication

1,111	2,222	3,333	4,444	5,555
x1,111	x2,222	x3,333	x4,444	x5,555
1,234,321	4,937,284	11,108,889	19,749,136	30,858,025

Explanation 4-digit multiplication

2,222
X2,222
4,937,284

1. 2x2=4
2. 2x2+2+2=8
3. 2x2+2x2+2x2=12 (Put down your 2 carry your 1)=2
4. 2x2+2x2+2x2+2+2+1=17 (Put down 7 carry your 1)=7
5. 2+2+2+2+2+2+1=13 (Put down 3 carry 1)=3
6. 2x2+2x2+1=9
7. 2x2=4
8. Final Answer: 4,937,284

3,333
X3,333
11,108,889

1. 3x3=9 (Put down 9)=9
2. 3x3+3x3=18 (Put down 8 carry 1)=8
3. 3x3+3x3+3x3+1=28 (put down 8 carry 2)=8
4. 3x3+3x3+3x3+3x3+2=38 (Put down 8 carry 3)=8
5. 3x3+3x3+3x3+3=30 (Put down 0 carry 3)=0 cont.
6. 3x3+3x3+3=21 (Put down1 carry 2)=1
7. 3x3+2=11 (Put down 11)=11
8. Final Answer: 11,108,889

The pure heart shall see God

Your progress in this walk will depend on your perseverance in your prayer life.

Put Answers
On these   1   2   3
Problems: 3333   2222   1111
                x 3333   x 2222   x 1111

choose the correct answer and place under the correct problem
1. 1,108,889
2. 4,937,284
3. 1,234,321

Practice: check: 2 times tables, here 1-12
2x1=2
2x2=4
2x3=6
2x4=8
2x5=10
2x6=12
2x8=16
2x9=18
2x10=20
2x11=22
2x12=24

PART VII

5 DIGIT MULTIPLICATION

5-digit Mental Multiplication

1.	2.	3.	4.	5.
11,111	12,112	13,121	14,121	15,121
x11,111	x 12,112	x 13,121	x 14,121	x 15,121
123,454,321	146700544	172,160,641	199,402,641	228,644,641

Practice;   11,111
             11,111
         123,454,321

Check your answer

11,111
11,111

1. 1x1=1
2. 1x1+1x1=2
3. 1x1+1x1+1x1=3
4. 1x1+1x1+1x1+1x1=4
5. 1x1+1x1+1x1+1x1+1x1=5 cont.
6. 1x1+1x1+1x1+1x1=4 cont.
7. 1x1=1x1=1x1=3
8. 1x1+1x1=2
9. 1x1=1
10. Final Answer: 123,454,321

Explanation question No. 2
5 digit multiplication:

  12,112   1. 2x2=4
x12,112   2. 1x2+2x1=4
146,700,544   3. 2x1+1x2+1x1=5 Cont.
                    4. 2x2+2x2+1x1+1x1=10
                    5. 1+1x2+1x2+1x2+2x1+1x1=10
                    6. 1+1x1+1x1+1x2+1x2=7
                    7. 1x1+1x1+2x2=6
                    8. 2x1+1x2=4
                    9. 1x1=1
                    10. Final Answer: 146,700,544

Explanation for 5 digits Cont.

13,121   1. 1x1=1
13,121   2. 1x2+1x2=4
172,160,641   3. 1x1+1x1+2x2=6. Cont.
                    4. 1x3+3x1+2x+1x2=10
                    5. 1+1x11x1+3x2+3x2+1 x 1=16
                    6. 1+1x2+1x2+3x1+1x3=11
                    7. 1+1x1+1x1+3x3=12
                    8. 1+ 3x1+1x3=7
                    9. 1x1=1
                    10. Final Answer: 172,160,641

PART VIII

6-7 DIGIT MULTIPLICATION

6-digit Mental Multiplication

	1.	2	3	4
	161,111	171,211	181,221	191,121
	x161,111	x171,211	x181,221	x191,121
	25,956,754,321	29,313,206,521	32,841,050,841	36,527,236,641

Explanation
6-digit multiplication

161,111
x161,111
25,956,754,321

Steps N0. 1
1. 1x1=1 (Put down 1)
2. 1x1+1x1=2 (Put down 2)
3. 1x1+1x1+1x1=3 (Put down 3)
4. 1x1+1x1+1x1+1x1=4 (Put down 4)
5. 1x6+6 x 1+1x1+1x1+1x1=15 (Put down 5 carry 1)
6. 1+1x2+1x2+3x1+1x3=11 (Put down 1 carry 1)
7. 1+1x1+1x1+3x3=12
8. 1+3x1+1x3=7
9. 1x1=1
10. 25,956,754,321

Six-Digit multiplication cont. No. 3

181,221
181,221
32,841,050,841

1. 1x1=1
2. 1x2+1x2=4
3. 1x2+2x1+2x2=8
4. 1x1+1x1+2x2+2x2=10
5. 1+1x8+8x1+2x1+1x2+1x2+2x2=25
6. 2+1x1+2x8+2x8+8x2+2x1+2x1=40
7. 4+2x1+1x2+1x8+1x8+1x1=41
8. 4+2x1+1x2+8x1+8x1=24
9. 2+1x1+1x1+64=68
10. 6+8 x 1+1 x 8=22
11. 1x 1+2=3 Final Answer:32,841,050,841

111,111	123,123	678,112	113,114	123,111
X111,111	x112,112	x234,512	x121,121	134,123
12,345,654,321	13,803,565,776	159,025,401,344	13,700,480,794	16,512,016,653

7-digit Mental Multiplication

1.	2.	3.
1,112,131	3,313,312	3,411,211
X1,211,211	x1,121,123	x2,113,221
1,347,025,300,641	3,714,630,289,376	7,208,642,720,631

Explanation 7 digit No.

1,112,131
X1,211,211
1,347,025,300,641

1. 1x1=1
2. 1x3+1 x 1=4
3. 1x1+2x1+3x1=6
4. 2x1+1x1+1x1+2x3=10 (Put down 0 carry 1)=0
5. 1+1x1+1x1 +2x1+1x3+1x2=10 (Put down 0 carry 1)=0
6. 1+1x1+2x1+2x1+1x3+2x2+1x1=13
7. 1+1x1+1x1+1x1+1x1+2x3+1x2+1x1+2x1=15 (Put down 5 carry 1)=5
8. 1+1x1+1x3+2x1+1x2+1x1+2x1=12 (Put down 2 carry1)=2 cont.
9. 1+2x1+1x1+1x1+2x2+1x1=10 (Put down 0 carry 1) =
10. 1+1x1+1x2+1x1+2x1=7
11. 1x1+1x1+1x2=4
12. 1x2+1 x 1=3
13. 1x1=1
    (Note: Answer, Left to right)

3,313,312
X1,121,123
3,714,630,289,376

1. 2x3=6
2. 1x3+2x2=7
3. 3x3+1x2+1x2=13 (Put down 3 carry 1)
4. 1+3x3+1 x 2+3x2+1 x 1=19 (Put down 9 carry 1)=9
5. 1+1x3+2x2+3x2+1x1+1x3=18 (Put down 8 carry1)=8
6. 3x3+1x2+1x2+3x1+3x1=22 (Put down 2 carry 2)=2
7. 2+3x3+2x1+3x2+1x1+1x1+2x3+3x1=30 Put down 0 carry 3)=0
8. 3+3x2+1 x 1+3x1+1x3+1 x 1+3x2=23 (Put down 3 carry 2)=3
9. 2+3x1+1x3+3x1+1x3+1x2=16 (Put down 6 carry 1)=6
10. 1+3x1+1x3+3x2+1 x 1=14 (Put down 4 carry 1)=4
11. 1+3x2+1x1+1x3=11 (Put down 1 carry 1)=1
12. 1+3x1+3x1=7
13. 1x3=3

Final Answer: 3,714,630,289,376 (Note: Answer Left to right)

PART IX

8-9 DIGIT
MULTIPLICATION

8-digit Mental Multiplication

12,341,555	55,555,55 5	55,555,555	55,432,122
x12,212,321	x12,112,3 21	x12,345,123	x11,111,111
150,719,031,299,155	672,906,715,493,155	685,840,159,808,265	615,912,460,507,542

5.   6.

55,555,555	28,112,312
11,111,111	21,221,122
617,283,938,271,605	596,574,802,654,064

Explanation Problem 5.
1. 1x5=5
2. 1x5+5x1=10
3. 1x5+1x5+1x5=16
4. 1+1x5+1x5+1x5+1x5=21
5. 1+1x5+1x5+1x5+1x5+1x5=27
6. 2+1x5+5x1+1x5+1x5+1x5+1x5=32
7. 3+1x5+5x1+1x5+1x5+1x5+1x5+1x5=3
8. 3+1x5+1x5+1x5+1x5+1x5+1x5+1x5+1x5=43
9. 4+1x5+1x5+1x5+1x5+1x5+1x5+1x5=39
10. 3+1x5+1x5+1x5+1x5+1x5+1x5=33
11. 3+5x1+5x1+1x5+1x5+1x5=28   Cont.
12. 2+1x5+1x5+1x5+1x5=22
13. 2+1x5+1x5+1x5=17
14. 1+1x5+1x5=11
15. 1+1x5=6

Final Answer: 61,283,938,271,605

9-digit Mental Multiplication

1	2.	3.	4.
611,234,212	211,345,233	123,412,121	112,233,444
X1111111111	x111111111	x111111111	x111111111
67,914,912,376,529	23,482,803,643,183,863	13,712,457,875,176,431	12,470,382,654,196,284

PART X

MYSTERY MULTIPLICATION: THE BEGINNING

An easier way to do do 9-digit 2's
Find the Greatest Common Factor for "2", it is " 4"
Use the series of numbers according to the number of digits. 1,2,3,4,5,6,7,8,9,8,7,6,5,4,3,2,1
Multiply
Add
Final Answer (backward)
Reversal
Final Answer(forward)
As follows:

222 222 222
222 222 222
49,382,715,950,617,284

4	4	4	4		4	4		4	4		4	4	4		4	4		4	4	4	GCF
1	2	3	4		5	6		7	8		9	8	7		6		5	4		3 2 1	Multiply

4 8 12 16 20 24 28 32 36 32 28 24 20 16 12 8 4
          1   1   2   2   3   3   3   3   3   2   2   1 1 Add
4 8 12 1 7 21 26 30 35 39 35 31 27 22 18 13 94 Answer Backward: 48, 271,605,951,728,394

Cont.

Reversal; Answer will be presented forward

4 8 12 16 20 24 28 32 36 32 28 24 20 16 12 84 (use from above)
    1 1 2 2 3 3 3 3 3 2 2 1 1 (add use from above)
4 9 13 18 22 27 31 35 39 35 30 26 21 17 12 84 Final Answer: 49,382,715,950,617,284 (forward)

Mystery Multiplication

3 3 3 3 3   Find Greatest Common Factor "21"
7 7 7 7 7   Use these multiple series of numbers: 1,2,3,4,5,4,3,2,1 Multiply by the Greatest Common
2,592,540,741   Factor

21	21	21	21	21	21	21	21	21 GCF
1	2	3	4	5	4	3	2	1 multiply
21	42	63	84	105	84	63	42	21 add first number to number multiplied
	2	4	6	9	11	9	7	4
21	44	67	90	114	95	72	49	25 Final Answer: (backward)

Cont.   1,470,452,952

Reversal

21	42	63	84	105	84	63	42	21 Add the above numbers in reverse
4	7	9	11	9	6	4	2	
25	49	72	95	114	90	67	44	21 Final Answer added: 2,592,540,741 forward

3 3 3 3
9 9 9 9
33,326,667

27	27	27	27	27	27	27 GCF
1	2	3	4	3	2	1
27	54	81	108	81	54	27
	2	5	8	11	9	6
27	56	86	116	92	63	33 Final Answer: (backward) 33,326,667 cont.

Reversal:

27	54	81	108	81	54	27 Add
6	9	11	8	5	2	
33	63	92	116	86	56	27 Final Answer:33,326,667 (forward)

(No Computer Needed)

Mysterious Multiplication cont.

2 2 2 2
9 9 9 9
22,217,778

| 98 |

Multiply

18	18	18	18	18	18	18	Greatest Common Factor
1	2	3	4	3	2	1	
18	36	54	72	54	36	18	Add
	1	3	5	7	6	4	
18	37	57	77	61	42	22	

Final Answer: 22,217,778 (choose 2nd number to add. The answer will be backward.

Reversal

18	36	54	72	54	36	18
4	6	7	5	3	1	
22	42	61	77	57	37	18

Final Answer: 22,217,778

Another Way

	1	3	5	7	6	4	Carry these numbers
18	18	18	18	18	18	18	Greatest Common Factor
1	2	3	4	3	2	1	Multiply
18	37	57	77	61	42	22	Final Answer backward:87,771,222

Reversal: 22,217,778

3 3 3 3
9 9 9 9
33,326,667

27	27	27	27	27	27	27	Greatest Common Factor
1	2	3	4	3	2	1	Multiple
27	54	81	108	81	54	27	Add
	2	5	8	11	9	6	
27	56	86	116	92	63	33	

Final Answer: 33,326,667

Reversal

27	54	81	108	81	54	27
6	9	11	8	5	2	Add
33	63	92	116	86	56	27

Final Answer:33,326,667

The toughest thing you can face is putting your own will under subjection to God.

Mystery Multiplication Cont.

```
  4  4  4  4
x 7  7  7  7
34,560,988
```

28	28	28	28	28	28	28	Multiply by GCF
1	2	3	4	3	2	1	
28	56	84	112	84	56	28	Add the first number result
	2	5	8	12	9	6	
28	58	89	120	96	65	34	34,560,988 (backward)

Reversal

28	56	84	112	84	56	28	Add
6	9	12	8	5	2		
34	65	96	120	89	58	28	Final Answer: 34,560,988

```
  5  5  5  5
x 7  7  7  7
43,201,235
```

Greatest Common Factor "35"

35	35	35	35	35	35	35	Multiply GCF
1	2	3	4	3	2	1	
35	70	105	140	105	70	35	Add
	3	7	11	15	12	8	
35	73	112	151	120	82	43	Final Answer:53,210,243 (backward)

Reversal

35	70	105	140	105	70	35	
8	12	15	11	7	3		Final Answer: 43,201,235 (forward)
43	82	120	151	112	73	35	

```
5   5    5    5    5
8   8    8    8    8
4,  9 3 8, 172,840
```

```
40   40    40 40 40    40    40    40 40 Multiply  GCF
 1    2     3  4  5      4     3     2  1
40   80    120 160 200  160  120  80 40 Add
      4      8   12  17   21   18 13  9
40  84     128 172 217 181 138  93 49(Final Answer backward)
```

Reversal

```
40  80   120    160    200 160 120    80    40
 9  13    18     21     17  12   8      4
49  93   138    181    217 172 128    84     40    Final Answer:  4,938,172,840
```

Duplicate Multiplication Example 49,382,715,950,617,284

```
222,222,222
222,222,222
49,382,715,950,617,284
```

Cont.
An Easier Way
Find the Greatest Common Factor = 4, 9,1 6, 25, 36, 49 64, 81

4. Examples: 9 digit numbers

(1)

	(2)	(3)
222,222,222	333,333,333	444,444,444
222,222,222	333,333,333	444, 444,444
49,382,715,950,617,284	111,111,110,888,888,889	197,530,863802,469,136

```
    555,555,555              666,666,666              777,777,777
    555,555,555              666,666,666              777,777,777
197,530,863,802,469,136  444,444,443,555,555,556  604,938,270,395,061,729

    888,888,888              999,999,999
    888,888,888              999,999,999
790,123,455,209,876,544  999,999,998,000,000,001
```

The other explanation to problem 4. (1)

1. 2x2=4
2. 2x2+2x2=8
3. 2x2+2x2+2x2=12
4. 1+2x2+2x2+2x2+2x2=17
5. 1+2x2+2x2+2x2+2x2+2x2=21
6. 2+2x2+2x2+2x2+2x2+2x2+2x2=26
7. 2+2x2+2x2+2x2+2x2+2x2+2x2+2x2=30
8. 3+2x2+2x2+2x2+2x2+2x2+2x2+2x2+2x2=35
9. 3+2x2+2x2+2x2+2x2+2x2+2x2+2x2+2x2=39
10. 3+2x2+2x2+2x2+2x2+2x2+2x2+2x2=35
11. 3+2x2+2x2+2x2+2x2+2x2+2x2+2x2=31
12. 3+2x2+2x2+2x2+2x2+2x2+2x2=27 Cont.
13. 2+2x2+2x2+2x2+2x2+2x2=22 cont.
14. 2+2x2+2x2+2x2+2x2=18 Cont.
15. 1+2x2+2x2+2x2=13
    1+2x2+2x2=9
    2x2=4
16. Final Answer: 49,382,715,950,617,284

MORE MYSTERY MULTIPLICATION

Please do the following: Use two numbers to multiply to get the Greatest Common Factor. (A number that all 4 numbers can be divided into evenly, no remainder,).

Once that number is established, limit yourself to 4 numbers to multiply, for a start.

To start;or find numbers for which you can be more comfortable. Multiply these numbers as you have previously observed beginning with the first digit on the right. Find the least common multiple:

Below, "16" is that number.

8 8 8 8
2 2 2 2
19,749,136

Step 1. Find the Greatest Common Factor, which will be "16."

2. Number to 7 digits as follows: 1,2,3,4,3,2,1

3. Multiply 2x8 to each of the four digits.

4. Begin at the right 2x4 digits cont.

5. Add the second number of each added number received as an answer.

6. This method will always elicit the right answer but it will be <u>backward</u>

16 16 16 16 16 16 16 GCF
1   2   3   4   3   2   1   Multiply
16 32 48 64 48 32 16
    1   3   5   6   5   3   Add
16 33   51 69 54 37 19   Final Answer:19,749,136 (Backward)

Cont.

Reversal

16 32 48 64 48 32 16
3   5   6   5   3   1
19 37 54 69 51 33 16 Final Answer:19,749,136

```
4   4   4   4
6   6   6   6
2 9,6 2 3,7 0 4
```

Mystery Multiplication Cont.

```
24    24  24    24     24   24   24 Greatest Common Factor for 4, and 6
 1     2   3     4      3    2    1 Multiply  Cont.
24    48  72    96     72   48   24
       2   5     7     10    8    5 Add
24    50  77   103     82   56   29 Final answer:29,623,704 (backward) Cont.
```

Reversal:

```
24  48  72  96  72  48  24
 5   8  10   7   5   2
29  5 6  82  103 77 50  24 Final Answer: 29,623,704
```

```
6  6  6  6
2  2  2  2
14,811,852
```

```
12  12  12  12  12  12   12 GCF
 1   2   3   4   3   2    1 Multiply
12  24  36  48  36  24   12
     1   2   3   5   4    2 Add
12 25  38  51  41  28   14 Final Answer 14,811,852 ( forward)
```

Reversal:

```
12  24  36   48  36  24  12
 2   4   5    3   2   1
14   2 8 4 1  5 1 38  25  12 Final Answer: 14,811,852
```

2, 2 2 2
7, 7 7 7
17,280,494

14	14	14	14	14	14	14	Greatest Common Factor
1	2	3	4	3	2	1	Multiply
14	28	42	56	42	28	14	
	1	2	4	6	4	3	Add
14	29	44	60	48	32	17	Final Answer: 17,280,494 Actual underlined number answer is backward)

Cont.
Reversal

14	28	42	56	42	28	14
3	4	6	4	2	1	
17	32	48	60	44	29	14 Final Answer:17,280,494

6, 6 6 6
6, 6 6 6
44,435,556

36	36	36	36	36	36	36 Greatest Common Factor
1	2	3	4	3	2	1 Multiply cont.
36	72	108	144	108	72	36
	3	7	11	15	12	8 Add
36	75	115	155	123	84	44 Final Answer: 65,553,444 (backward)

Cont.

Reversal:

36	72	108	144	108	72	36
8	12	15	11	7	3	
44	84	123	155	115	75	36 Final Answer:44,435,556

Mystery Multiplication Cont.
A new way and a new way to do number series multiplication
4 digits

3,333
2,222
7,405,926

How to get the answer: find the <u>Greatest Common Factor which</u> would be "6."

```
6   6   6   6   6   6   6
1   2   3   4   3   2   1  Multiply
6  12  18  24  18  12   6
          1   1   2   2   1 Add
6  12  19  25  20  14   7 Answer  is underlined: 6,295,047 (Presented backward)
```

Cont.

Reversal

```
6  12 18  24  18 12 6 Add
1   2  2   1   1
7  14 20  25  19 12 6   Final Answer: 7,405,926
```

Example (2)

```
4    4    4    4
5    5    5    5          Find (Greatest Common Factor)=20
24,6 8 6,4 2 0           (What number 4,and 5 can divide into evenly)
```
Cont.

Pattern:
```
20   20    20    20   20   20   20 Greatest Common Factor "20"
 1    2     3     4    3    2    1 Multiply   (use this series of numbers 1,2,3,4,3,2,1 to multiply)
20   40    60    80   60   40   20  Put first number beneath the second number and
      2     4     6    8    6    4       Add (use this series of numbers 2,4,6,8,6,4)
20   42    64    86   68   46   24  Answer will be backward : 24,686,420 to above problem
```

Cont.
Reversal

```
20   40    60    80   60   40   20
 4    6     8     6    4    2
24   46    68    86   64   42   20 Final Answer:  24,686,420
```

Mystery Multiplication Cont.

<u>8888 8888 Multiply</u>
78,996,544

64	64	64	64	64	64	64	Greatest Common Factor
<u>1</u>	<u>2</u>	<u>3</u>	<u>4</u>	<u>3</u>	<u>2</u>	<u>1 Multiply</u>	
64	128	192	256	192	128	64	
	<u>6</u>	<u>13</u>	<u>20</u>	<u>27</u>	<u>21</u>	<u>14 Add</u>	
6<u>4</u>	13<u>4</u>	20<u>5</u>	27<u>6</u>	21<u>9</u>	14<u>9</u>	<u>78</u>	Your answer will be backward: 78,996,544

Reversal

64	128	192	256	192	128	64
<u>14</u>	<u>21</u>	<u>27</u>	<u>20</u>	<u>13</u>	<u>6</u>	
<u>78</u>	14<u>9</u>	21<u>9</u>	27<u>6</u>	20<u>5</u>	13<u>4</u>	64 Final Answer: 78,996,544

   2,222
  <u>4,444</u>
9,874,568

8	8	8	8	8	8	8 GCF
<u>1</u>	<u>2</u>	<u>3</u>	<u>4</u>	<u>3</u>	<u>2</u>	<u>1 Multiply</u>
8	16	24	32	24	16	8
		<u>1</u>	<u>2</u>	<u>3</u>	<u>2</u>	<u>1 Add</u>
<u>8</u>	1<u>6</u>	2<u>5</u>	<u>4</u>	2<u>7</u>	1<u>8</u>	<u>9</u> Final answer: 8,654,789 (backward)

8	16	24	32	24	16	8
<u>1</u>	<u>2</u>	<u>3</u>	<u>2</u>	<u>1</u>		
<u>9</u>	1<u>8</u>	<u>27</u>	3<u>4</u>	2<u>5</u>	16	<u>8</u> Final Answer:9,874,568 Reversal (forward)

Mystery Multiplication

2222
<u>4444</u>
9,874,568

    1 2 3 2 1
8 8 8 8 8 8 8
<u>1 2 3 4 3 2 1</u>
8 6 5 4 7 8 9 (backward)

9 , 8 7 4 , 5 6 8 (forward)

5 5 5 5
6 6 6 6
37,029,630

30 30   30 30 30   30 30 GCF
1   2   3   4   3   2   1
30 60   90 120 90 60 30
       3   6   9   12 10   7
30 63 96 129 102 70 37(backward)

37,029,630 (forward)

5555
7777
4 3 , 2 0 1 , 2 3 5
       3   7 10
35 35 35 35   35   35 35
1   2   3   4   3   2   1
35 70   105 140 105 70 35
      3   7 12 15 12   8
35 73   112 152 120 82 43 43,201,235(no computer)

4,444
5,555
24,686,420 Cont.

20 20 20 20 20 20 20
1   2   3 4   3   2   1
20 40 60 80 60 40 20
      4 2 4   6   8   6   4
20 42 64 86 68 46 24 (forward)24,686,420)

3,333
4,444
14,811,852 cont.

```
12  12  12  12  12   12   12 GCF
 1   2   3   4   3    2    1
12  24  36  48  36  24   12
     1   2   3   5    4    2
 12 25  38  51   41  28  14(backward) 14,811,852 (forward)
```

No computer needed

More Examples: 9-digits

211,345,233
x511,231,234
108,046,284,266,607,522

1. 3x4=12 (Put 2 down carry 1)=2
2. 1+3x4+3x3=22 (Put down 2 carry 2)=2
3. 2+2x4+2x3+3x3=25 (Put down 5 carry 2)=5
4. 2+5x4+1x3+2x3+2x3=37 (Put down 7 carry 3)=7
5. 3+4x4 +3x3+5x3+1x3+2x2=50 (Put down 0 carry 5)=0
6. 5+ 3x4+ 2x3 4x3+ 3x3+5x2+1x2=56 (Put down 6 carry 5)=6
7. 5+1x4 +1x3+ 2x3 +3x3+2x4+2x5+ 1x2+ 46 (Put down 6 carry 4)=6
8. 4+1x4+1x3+1x3+1x3+1x3+1x3+3x2+2x2+1x4+3x5=46 (Put down 6 carry 4)=6
9. 4+ 2x4+5x3+1x3+1x3+1x2+1x2+1x3+2x5+4x3=62 (Put down 2 carry 6)=2
10. 6+2x3+5x3+1x2+1x2+1x5+1x1+3x3+2x4=54 (Put down 4 carry 5)=4
11. 5+2x2+2x5+2x5+1x1+5x1+3x1+4x1+3x2=38
12. 3+2x1+5x5+3x1+3x1+1x4+1x2+1x3=42 (Put down 2 carry 4)=2
13. 4+2x3+4x5+1 x 2+3x1+1 x 1=36 (Put down 6 carry 3)=6
14. 3+ 2x2+ 5x3+ 1x1+1x1=24 (Put down 4 carry 2)=4
15. 2+2x1+5x1+1 x 1=10 (Put down 0 carry 1)=0
16. 1+ 2x1+1 x 5=8 (Put down 8)
17. 2x5=10 (put down 10)=10   Final Answer: 108,046,284,266,607,522

Review

111,111,111
X111,111,111
12,345,678,987,654,321

9-digit multiplication steps

1. 1x1=1
2. 1x1+1x1=2
3. 1x1+1x1+1x1=3
4. 1x1+1x1+1x1+1x1=4
5. 1x1+1x1+1x1+1x1+1x1=5

Cont. 6. 1x1+1x1+1x1+1x1+1x1+1x1=6
7. 1x1+1x1+1x1+1x1+1x1+1x1+1x1=7
8. 1x1+1x1+1x1+1x1+1x1+1x1+1x1+1x1=8
9. 1x1+1x1+1x1+1x1+1x1+1x1+1x1+1x1+1x1=9
10. 1x1+1x1+1x1+1x1+1x1+1x1+1x1+1x1=8 Cont.
11. 1x1+1x1+1x1+1x1+1x1+1x1+1x1=7
12. 1x1+1x1+1x1+1x1+1x1+1x1=6
13. 1x1+1x1+1x1+1x1+1x1=5
14. 1x1+1x1+1x1+1x1=4
15. 1x1+1x1+1x1=3 Cont.
16. 1x1+1x1=2
17. =1
Final Answer: 12,345,678,987,654,321

Mystery Multiplication Advance

How to: Heretofore you have had the the following formula: 1 2 3 4 3 2 1
This formula depends on the number used in your problem. Use 7 numbers i.e. 1 2 3 4 3 2 1
If you use "5" numbers, use as follows: 1 2 3 4 5 4 3 2 1
If you use 6 numbers, use as follows: 1 2 3 4 5 6 5 4 3 2 1 etc.

Mystery Multiplication cont. (Advance Method)

Let's see our first example of an extension: 66,666   5 digits
                                                                          22,222
                                                     1,481,451,852 Cont.

12	12	12	12	12	12	12	12	12	Greatest Common Factor
1	2	3	4	5	4	3	2	1	Multiply (Use these series of numbers: 1,2,3,4,5,4,3,2,1)
12	24	36	48	60	48	36	24	12	
	1	2	3	5	6	5	4	2	Add
12	25	38	51	65	54	41	28	14	Answer:(258,154,184,1 the answer is backward)

Cont.

Reversal

12 24 36 48 60 48 36 24 12
 2 4 5 6 5 3 2 1
14 28 41 54 65 51 38 25 12 Final Answer:1,481,451,852

PART XI

PRACTICE SHEETS FOR THE MULTIPLICATION SERIES CONTINUE

Practice Do an example of Mental Multiplication

Practice: Do 3 time tables 1-12

Review

Check for 3 time tables: 3x12=_____3x9=_____
3x1=3
3x2=6
3x3+9
3x4=12
3x5=15
3x6=18
3x7=21
3x8=24
3x9=27
3x10=30
3x11=33
3x12=36

Practice: Do 2 time tables 1-12 see page 35-70   2x7=_____2x8=_____

2x1=2
2x2=4
2x3=6
2x4=8
2x5=10
2x6=12
2x7=14
2x8=16
2x9=18
2x10=20
2x11=22
2x12=24

Practice: Do 4 time tables 1-12 see pg 35-70

Review for Accuracy

4x1=4   4x5=_____4x9=_____4x6=_____4x7=_____
4x2=8
4x3=12
4x4=16
4x5=20
4x6=24
4x7=28
4x8=32
4x9=36
4x10=40
4x11=44
4x12=48

Practice Do 5 time tables

5 Check   Review for Accuracy   5x6=_____5x9=_____5x7=_____
5x1=5
5x2=10
5x3=15
5x4=20
5x5=25
5x6=30
5x7=35
5x8=40
5x9=45
5x10=50
5x11=55
5x12=60

Check: Do 6 time tables 1-12

6 check   Review for Accuracy   6x11=_____ 6x2=_____ 6x9=_____
6x1=6
6x2=12
6x3=18
6x4=24
6x5=30
6x6=36
6x7=42
6x8=48
6x9=54
6x10=60
6x11=66
6x12=72

Do 7 time tables

check: 7 time tables-1-12

7 Check   Review for Accuracy   7x12=_____7x11=_____7x5=_____7x9=_____
7x1=7
7x2=14
7x3=21
7x4=28
7x5=35
7x6=42
7x7=49
7x8=56
7x9=63
7x10=70
7x11=77
7x12=84

Practice 8 timetables

Check 8 times tables   8 x 3=_____8x9=_____8x7=_____
8x1=8
8x2=16
8x3=24
8x4=32
8x5=40
8x6=48
8x7=56
8x8=64
8x9=72
8x10=80
8x11=88
8x12=96

Practice: Do 9 time tables

9x1=9 Review for Accuracy 9x3=_____9x5=_____9x4=_____9x8=_____

9x2=18
9x3=27
9x4=36
9x5=45
9x6=54
9x7=63
9x8=72
9x9=81
9x10=90
9x11=99
9x12=108

Practice: Do 11 time tables

Check time tables for 11. Review for Accuracy 11x11=_____

11x12_____11x5=_____11x6=_____

11x1=11
11x2=22
11x3=33
11x4=44
11x5=55
11x6=66
11x7=77
11x8=88
11x9=99
11x10=110
11x11=121
11x12=132

Review

2 2 2 2 2
3 3 3 3 3
7 4 0 ,7 2 5 ,9 2 6

Formula: Mystery Multiplication

1. Find GCF
2. Multiply
3. Add
4. Use the first number in the set for the answer
5. The Answer will be backward in presentation

<u>: Multiply Note the order of these numbers</u>

6   12   18   24   30   24   18   12   6
        1   1   2   3   2   2   1   Add <u>every other first number in the answer</u>
<u>6</u> 12 <u>19</u> 25 <u>32</u> 27 20 14 <u>7</u>   Answer: 7 4 0 ,7 2 5 ,9 2 6 (every other number)
(Do not add the first two numbers) Note: Numbers are backward in the answer

Reversal

6 12 18 24 30 24 18 12 6
<u>1   2   2   3   2   1   1</u>
<u>7</u> 1 <u>4</u> 2 <u>0</u> 2 <u>73</u> 22 <u>5</u> 19 12 <u>6</u> Final Answer: 740,725,926

Practice 36 time tables

Check 36 timetables   36x2=_____ 36x5=_____ 36x7=_____ 36x8=_____

36 x 1=36
36x2=72
36x3=108
36x4=144
36x5=180
36x6=216
36x7=252
36x8=288
36x9=324
36x10=360
36x11=396
36x12=432

```
5   5   5   5
6   6   6   6
37,029,630
```

Formula: Mystery Multiplication

1. Find Greatest Common Factor: 30
2. Multiply
3. Add
4. Use the first number in the set for the answer
5. The Answer will be backward in presentation

```
30    30    30    30    30    30    30 Greatest Common Factor
1     2     3     4     3     2     1 Multiply
30    60    90    120   90    60    30
      3     6     9     12    10    7   Add (do not add first number)
30    63    96    129   102   70    37 Answer will be backward:  ( every other number) 37,029,630
```

Reversal

```
30   60    90    120   90         60    30
7    10    12    9     6          3         Final Answer: 37,029,630
37   70    102   229   96         63    30
```

Practice :write 13 time tables 1-12 see pg 35-70

Check 13 timetables Review for Accuracy 13x5=_____ 13x6=_____ 13x9=_____

13x1=13
13x2=26
13x3=39
13x4=52
13x5=65
13x6=78
13x7=91
13x8=104
13x9=117
13x10=130
13x11=143
13x12=156

Practice: Do 14 time tables

*The life of God within you, enables you to begin
a journey to the greatest victory.*

Check 14 timetables Review for Accuracy 14x4=_____ 14x5=_____ 14x7=_____
14x10=_____

14x1=14
14x2=28
14x3=42
14x4=56
14x5=70
14x6=84
14x7=98
14x8=112
14x9=126
14x10=140
14x11=154
14x12=168

More Mystery Multiplication

```
3   3   3   3
4   4   4   4
14,811,852
```

1. Find the Greatest Common Factor: 1
2. Multiply
3. Add
4. Use the first number in the set for the answer
5. The Answer will be backward in presentation

```
12  12  12  12  12  12   12
 1   2   3   4   3   2    1  Multiply
12  24  36  48  36  24   12
     1   2   3   5   4    2  Add
12  25  38  51  41   28  14  Answer: 14,811,852  cont.
```

Reversal

```
12  24  36  48  36  24  12
 2   4   5   3   2   1
14  28  41  51  38  25  12  Final Answer: 14, 811,852
```

Extensions
```
33333
22222
740,725,926
```

Extension Cont.
```
33,333
22,222
740,725,926
```

```
6   6   6   6   6   6   6   6   6  Greatest Common Factor
1   2   3   4   5   4   3   2   1  Multiply
6   12  18  24  30  24  18  12  6
         1   1   2   3   2   2   1  Add
6   12  19  25  32  27  20  14  7  Answer: 740,725,926
```

Reversal

```
6  12  18  24  30  24  18  12  6
1   2   2   3   2   1   1
7  14  20  27  32  25  19  12  6  Final Answer: 740,725,926
```

```
6   6   6   6
7   7   7   7
51,841,482
```

Explanation:
1. Find Greatest Common Factor: 42
2. Multiply
3. Add
4. Use the second number in the set for the answer
5. The Answer will be backward in presentation

Mystery Multiplication Cont.

```
42    42    42    42    42    42   42
1     2     3     4     3     2    1  Multiply
42    84   126   168   126    84   42
       4     8    13    18    14  9 Add
42    88   134   181   144    98  51   Answer:51,841,482
```

Reversal

```
42  84  126  168 126   84   42
 9  14   18   13   8    4
51  98  144  181 134   88   42
```

```
4,444
5,555
2,468,642   Cont.
```

```
20    20    20   20    20    20   20 Least Common Multiple
1     2     3    4     3     2    1 Multiply
20    40    60   80    60    40   20
       2     4    6     8     6    4 Add
20    42    64   86    68         46   24 Answer: 2,468,642( Reversal)
```

5,555
<u>5,555</u>
30,858,025

25	25	25	25	25	25	25 Greatest Common Factor
<u>1</u>	<u>2</u>	<u>3</u>	<u>4</u>	<u>3</u>	<u>2</u>	<u>1 Multiply</u>
25	50	75	100	75	50	25
	<u>2</u>	<u>5</u>	<u>8</u>	<u>10</u>	<u>8</u>	<u>5 Add</u>
<u>25</u>	<u>52</u>	<u>80</u>	<u>108</u>	<u>85</u>	<u>58</u>	<u>30</u> Answer:52,085,830 (backward)

Reversal

25	50	75	100	75	50	25
<u>5</u>	<u>8</u>	<u>10</u>	<u>8</u>	<u>5</u>	<u>2</u>	
<u>30</u>	<u>58</u>	<u>85</u>	<u>108</u>	<u>80</u>	<u>52</u>	<u>25</u> Final Answer:30,858,025

Your Grace is predicated on nothing other than love.

You will never fail if you give God all the honor, praise and glory

Mystery Multiplication

6,666
6,666
44,435,556

36	36	36	36	36	36	36 Greatest Common Factor
1	2	3	4	3	2	1 Multiply
36	72	108	144	108	72	36
	3	7	11	15	12	8 Add
36	75	115	155	123	84	44 Answer:65,553,444 (backward)

Cont.
Reversal

36	72	108	144	108	72	36
8	12	15	11	7	3	
44	84	123	155	115	75	36 Final Answer: 44,435,556

7,777
7,777
60,481,729

49	49	49	49	49	49	49 Greatest Common Factor
1	2	3	4	3	2	1 Multiply
49	98	147	196	147	98	49 Add
	4	10	15	21	16	11
49	102	157	211	168	114	60 Answer: 60,481,729 (Reversal)(shortcut)

7,777
8,888
69,121,976

56	56	56	56	56	56	56 Greatest Common Factor
1	2	3	4	3	2	1 Multiply
56	112	168	224	168	112	56
	5	11	17	24	19	13 Add
56	117	179	241	192	131	69 Answer:69,121,976 (Reversal)

Reversal

56	112	168	224	168	112	56
13	19	24	17	11	5	
69	121	192	241	179	117	56

The touch of God goes a long way

8,888
8,888
78,996,544

Mystery Multiplication

64	64	64	64	64	64	64 Greatest Common Factor
1	2	3	4	3	2	1 Multiply
64	128	192	256	192	128	64
	6	13	20	27	21	14 Add
64	134	205	276	219	149	78 Answer: 78,996,544 (Reversal)

Reversal

64	128	192	256	192	128	64
14	21	27	20	13	6	
78	149	219	276	205	134	64 Final Answer: 78,996,544

8888
9999
88,871,112

72	72	72	72	72	72	72 Greatest Common Factor
1	2	3	4	3	2	1 Multiply
72	144	216	288	216	144	72
	7	15	23	31	24	16 Add
72	151	231	311	247	168	88 Answer:88,871,112(backward)

Reversal

72	144	216	288	216	144	72
16	24	31	23	15	7	
88	168	247	311	231	151	72 Final Answer: 88,871,112

Your peace will depend on emptying yourself to our Lord.

You don't fear man? Fear God.

The comparison between the mind of Christ and the flesh, is beyond human comprehension.

Check 37 time tables Review for accuracy 37X12=_____ 37x10=_____. 37x6=_____

37x1=37
37x2=74
37x3=111
37x4=148
37x5=185
37x6=222
37x7=259
37x8=296
37x9=333
37x10=370
37x11=407
37x12=444

Practice: write 14 time tables 1-12

14 time tables check 14x12=_____14x8=_____14x8=_____14x10=_____

14x1=14
14x2=28
14x3=42
14x4=56
14x5=70
14x6=84
14x7=98
14x8=112
14x9=126
14x10=140
14x11=154
14x12=168

Practice: write 16 time tables

16 tables check Review for Accuracy 16x12=_____ 16x7=_____ 16x8=_____
16x5_____

16x1=16
16x2=32
16x3=48
16x4=64
16x5=80
16x6=96
16x7=112
16x8=128
16x9=144
16x10=160
16x11=176
16x12=192

Practice: write 20 time tables

20 times table check 20x3=_____20x5=_____20x6=_____

20x1=20
20x2=40
20x3=60
20x4=80
20x5=100
20x6= 120
20x7=140
20x8=160
20x9=180
20x10=200
20x11=220
20x12=240

Practice: write 24 time tables

24 time tables check Review for Accuracy 24x7=_____ 24x8=_____ 24x9=_____
24x6=_____

24x1=24
24x2=48
24x3=72
24x4=96
24x5=120
24x6=144
24x7=168
24x8=192
24x9=216
24x10=240
24x11=264
24x12=288

Practice: write 30 time tables

30 time table check Review for Accuracy 12x30=_____ 30x9=_____
30x6=_____ 30x4=_____

1x30=30
2x30=60
3x30=90
4x30=120
5x30=150
6x30=180
7x30=210
8x30=240
9x30=270
10x30=300
11x30=330
12x30=360

Mystery Multiplication

4,444
4,444
19,749,136

16	16	16	16	16	16	16	Greatest Common Factor
1	2	3	4	3	2	1	Multiply
16	32	48	64	48	32	16	
	1	3	5	6	5	3	Add
16	33	51	69	54	37	19	Answer:63,194,719 (backward)

Cont.

Reversal

16	32	48	64	48	32	16	
3	5	6	5	3	1		
19	37	54	69	51	33	16	Final Answer: 19,749,136

Check 40 times tables Review for Accuracy 40x4=_____ 40x9=_____40x5=_____

40x1=40
40x2=80
40x3=120
40x4=160
40x5=200
40x6=240
40x7=280
40x8=320
40x9=360
40x10=400
40x11=440
40x12=480

Check 25 time tables 1-12 Review for Accuracy 25x6=_____25x7=_____25x8=_____

25x1=25
25x2=50
25x3=75
25x4=100
25x5=125
25x6=150
25x7=175
25x8=200
25x9=225
25x10=250
25x11=275
25x12=300

9999
9999
99,980,001

81	81	81	81	81	81	81	Greatest Common Factor
1	2	3	4	3	2	1	Multiply
81	162	243	324	243	162	81	
	8	17	26	35	27	18	Add
81	170	260	350	278	189	99	Answer: 10,008,999

Reversal

81	162	243	324	243	162	81	
18	27	35	26	17	8		
99	189	278	350	260	170	81	Final Answer: 99,980,001

Practice: write 10 time tables

10 times tables check 10x10=_____10x11=_____10x12=_____10x6=_____

10x1=10
10x2=20
10x3=30
10x4=40
10x5=50
10x6=60
10x7=70
10x8=80
10x9=90
10x10=100
10 x 11=110
10x12=120

Practice: write 12 time tables

12x1=12 Review for Accuracy 12x6=_____12x7=_____12x8=_____
12x5=_____

12x2=24
12x3=36
12x4=48
12x5=60
12x6=72
12x7=84
12x8=96
12x9=108
12x10=120
12x11=132
12x12=144

Write 5 time tables

Check 5 times tables Review for Accuracy 5x12=_____5x11=_____5x7=_____
5x8=_____

5x1= 5
5x2=10
5x3 =15
5x4 =20
5x5 =25
5x6=30
5x7=35
5x8=40
5x9=45
5x10=50
5x11=55
5x12=60

Practice: write 4 time tables

4x1=4 Review for Accuracy 4x8=_____ 4x9=_____ 4x7=_____ 4x12=_____

4x2=8

4x3=12

4x4=16

4x5=20

4x6=24

4x7=28

4x8=32

4x9=36

4x10=40

4x11=44

4x12=48

Practice: write 8 time tables

8x1=8 Review for Accuracy 8x4=_____ 8x7=_____ 8x5=_____ 8x9=_____
8x2=16
8x3=24
8x4=32
8x5=40
8x6=48
8x7=56
8x8=64
8x9=72
8x10=80
8x11=88
8x12=96

Extensions Continued (5 digits)

  44,444
  66,666
2,962,903,704

24	24	24	24	24	24	24	24	24 Greatest Common Factor(24)
1	2	3	4	5	4	3	2	1 Multiply
24	48	72	96	120	96	72	48	24
	2	5	7	10	13	10	8	5 Add right digit the answer to the multiplication keep "29"
24	50	77	103	130	109	82	56	29 Answer:407,309,262 answer will be backward.

Reversal

24 48 72 96 120 96 72 48 24
 5 8 10 13 10 7 5 2
29 56 82 109 130 103 77 50 24 Final Answer:2,962,903,704

```
3 3 3 3 3
9 9 9 9 9
```
3,333,266,667

27	27	27	27	27	27	27	27	27 Greatest Common Factor
1	2	3	4	5	4	3	2	1 Multiply
27	54	81	108	135	108	81	54	27
	2	5	8	11	14	12	9	6 Add
27	56	86	116	146	122	93	63	33 Answer:7,666,623,333 (backward)

Reversal

27	54	81	108	135	108		81	54	27 Add
6	9	12	14	11	8		5	2	
33	63	93	122	146	116		86	56	27 Final Answer: 3,333,266,667

Check 21 timetables Review for Accuracy 21x12=_____21x6=_____21x10_____
21x11=_____

21x1=21
21x2=42
21x3=63
21x4=84
21x5=105
21x6=126
21x7=147
21x8=168
21x9=189
21x10=210
21x11=231
21x12=252

Practice: write 31 time tables

Practice: write 31 time
Check Review for Accuracy 31x12=_____ 31x6=_____ 31x7=_____ 31x8=_____

31x1=31
31x2=62
31x3=93
31x4=124
31x5=155
31x6=186
31x7=217
31x8=248
31x9=279
31x10=310
31x11=341
31x12=372

Your expertise in multiplication comes through practice for mastery

Check: write 16 times tables 1-12 Review for Accuracy 16x8=_____ 16x5=_____
16x12=_____

16x1=16
16x2=32
16x3=48
16x4=64
16x5=80
16x6=96
16x7=112
16x8=128
16x9=144
16x10=160
16x11=176
16x12=192

Check 32 time tables Review for Accuracy 32x4=_____ 32x6=_____32x9=_____
32x10=_____

32x1=32
32x2=64
32x3=96
32x4=128
32x5=150
32x6=182
32x7=214
32x8=246
32x9=278
32x10=320
32x11=352
32x12=384

Practice: write 17 time tables

Check 17 time tables Review for Accuracy 17x4=_____17x6=_____17x5=_____
17x7=_____

17x1=17
17x2=34
17x3=51
17x4=68
17x5=85
17x6=102
17x7=119
17x8=136
17x9=153
17x10=170
17x11=187
17x12=204

Practice: write 19 time tables

Check write 19 time tables Review for Accuracy 19x4=_____ 19x7=_____ 19x5=_____
19x9=_____

19x1=19
19x2=38
19x3=57
19x4=76
19x5=95
19x6=114
19x7=133
19x8=152
19x9=171
19x10=190
19x11=209
19x12=228

Practice: write 21 time tables

Check: 21 time tables 21x5=_____21x7=_____21x9_____21x8=_____
21x10=_____

21x1=21
21x2=42
21x3=63
21x4=84
21x5=105
21x6=126
21x7=147
21x8=168
21x9=189
21x10=210
21x11=231
21x12=252

Check 22 timetables Review for Accuracy 22x7=_____22x5=_____22x8=_____
22x10=_____

22x1=22
22x2=44
22x3=66
22x4=88
22x5=110
22x6=132
22x7=154
22x8=176
22x9=198
22x10=220
22x11=242
22x12=264

Do 23 timetables

Check 23 times tables tablesReview for Accuracy 23x5=_____23x8=_____
23x7=_____23x4+_____

23x1=24
23x2=46
23x3=69
23x4=92
23x5=115
23x6=138
23x7=161
23x8=184
23x9=207
23x10=230
23x11=253
23x12=276

Check: 27 timetables Review for Accuracy 27x9=_____27x6=_____27x5=_____
27x8=_____

27x1=27
27x2=54
27x3=81
27x4=108
27x5=135
27x6=162
27x7=189
27x8=216
27x9=243
27x10=270
27x11=297
27x12=324

7,777,777
7,777,777
60,493,815,061,729

49	49	49	49	49	49	49	49	49	49	49	49	49 GCF
1	2	3	4	5	6	7	6	5	4	3	2	1
49	98	147	196	245	294	343	294	245	196	147	98	49 Add
	4	10	15	21	26	32	37	33	27	22	16	11
49	102	157	211	266	3 2 0		37 5	33 1	27 8	223	169	114 6 0 Answer: 92,716,051,839,460

Reversal

49	98	147	196	245	294	343	294	245	196	147	98	49 Add
11	16	22	27	33	37	32	26	21	15	10	4	
60	114	169	223	378	341	375	320	266	211	157	102	49 Answer:60,493,815,061,729

Check: 24 time tables Review for Accuracy 24x6=_____24x9=_____24x10=_____
24x7=_____

24x1=24
24x2=48
24x3=72
24x4=96
24x5=120
24x6=144
24x7=168
24x8=192
24x9=216
24x10=240
24x11=264
24x12=288

PART XII

THE HARDEST PROBLEMS IN THIS BOOK

These problems take concentration and study to achieve mastery

Examples: The Shortcut (Can you figure this out?) It took me a while.

101,010
x   2,222
224,444,220

Steps
1. 2x0=0
2. 2x1=2+2x0=2
3. 2x0+2x0+2 x 1=2
4. 2x1+2x0+2x0+2 x 1=4
5. 2x1+2x0+2x0=2 x 1=4
6. 2x1+2x0+2x0+2 x 1=4 Cont.
7. 2x1+2x0+2x0+2 x 1=4
8. 2x1+2+0=2
9. 2x1=2
Answer: 224,444,220

111,111
x   2222
246,888,642

Steps
1. 2x1=2
2. 2x1+2x1=4
3. 2x1+2x1+2x1=6   Cont.
4. 2x1+2x1+2x1+2x1=8
5. 2x1+2x1+2x1+2x1=8
6. 2x1+2x1+2x1+2x1=8
7. 2x1+2x1+2x1=6
8. 2x1+2x1=4
9. 2x1= 2
Answer: 246,888,642

```
  222,222          1.  2x2=4
x   2,222          2.  2x2+2x2=8
493,777,284        3.  2x2+2x2+2x2=12
                   4.  1+2x2+2x2+2x2+2x2=17
                   5.  1+1+2x2+2x2+2x2+2x2=17
                   6.  1+2x2+2x2+2x2+2x2=17
                   7.  1+2x2+2x2+2x2=13
                   8.  1+2x2+2x2=9
                   9.  2x2=4
                   Answer: 493,777,284
```

```
  333,333          1.  2x3=6
x   2,222          2.  2x3+2x3=12
740,665,926        3.  1+2x3+2x3+2x3=19
                   4.  1+2x3+2x3+2x3+2x3=25
                   5.  2+2x3+2x3+2x3+2x3=26
                   6.  2+2x3+2x3+2x3+2x3=26
                   7.  2+2x3+2x3+2x3=20
                   8.  2+2x3+2x3=14
                   9.  1+2 x 3=7
                   Answer: 740,665,926
```

```
  444,444          1.  2x4=8
x 2,222            2.  2x4+2x4=16
987,554,568        3.  1+2x4+2x4+2x4=25
                   4.  2+2x4+2x4+2x4+2x4=34
                   5.  3+2x4+2x4+2x4+2x4=35
                   6.  3+2x4+2x4+2x4+2x4=35
                   7.  3+2x4+2x4+2x4=27
                   8.  2+2x4+2x4=18
                   9.  1+2x4=9 R
                   Answer: 987,554,568
```

24 time tables

Please write: Review for Accuracy 24x5=_____ 24x7=_____ 24x8=_____ 24x9=_____

24 time tables check

24x1=24
24x2=48
24x3=72
24x4=96
24x5=120
24x6=144
24x7=168
24x8=192
24x9=216
24x10=240
24x11=264
24x12=288

Practice: Do one example of Mystery Multiplication

Practice: Do one example of Mental Multiplication

Practice: Do 18 times tables 1-12

Check 18 timetables Review for Accuracy 24x18=_____24x10=_____24x6=_____
24x9=_____

18x1=18
18x2=36
18x3=54
18x4=72
18x5=90
18x6=108
18x7=126
18x8=144
18x9=162
18x10=180
18x11=198
18x12=216

Practice: Do 39 timetables Review for Accuracy 39x10=_____39x5=_____
39x9=_____

39x1=39
39x2=78
39x3=117
39x4=156
39x5=195
39x6=234
39x7=273
39x8=312
39x9=351
39x10=390
39x11=429
39x12=468

Practice; Do 40 time tables 1-12

Check: Do 40 time tables 40x2=_____40x4=_____40x5=_____40x10=_____
40x9=_____

40x1=40
40x2=80
40x3=120
40x4=160
40x5=200
40x6=240
40x7=280
40x8=320
40x9=360
40x10=400
40x11=440
40x12=480

Practice: Do 41 time tables 1-12

Check: 41 time tables 41x4=_____ 41x6=_____ 41x7=_____ 41x8=_____
41x9=_____

41x1=41
41x2=82
41x3=123
41x4=164
41x5=205
41x6=246
41x7=287
41x8=328
41x9=369
41x10=410
41x11=451
41x12=492

Practice: Do 34 time tables 1-12

Check: Do 34 time tables 34x5=_____ 34x6=_____ 34x9=_____ 34x10=_____

34x1=34
34x2=68
34x3=102
34x4=136
34x5=170
34x6=204
34x7=238
34x8=272
34x9=306
34x10=340
34x11=374
34x12=408

Do 50 time tables 1-12

Check: Do 50 time tables 1-12

Review for Accuracy 50x4=_____   50x2= 50x3   50x9=_____

50x1=50
50x2=100
50x3=150
50x4=200
50x5=250
50x6=300
50x7=350
50 x 8=400
50x9=450
50x10=500
50x11=550
50x12=600

Practice: Study this problem:

222   1. 2x2=4
x222   2. 2x2+2x2=8
             3. 2x2+2x2+2x2=12
             4. 1+2x2+2x2=9   Cont.
             5. 2x2=4
             Final Answer: 49,284

Do this problem

222   1.
X222   2.
             3.
             4.
             5.

Practice: Study this problem:

333
<u>333</u>

1. 3x3=9
2. 3x3+3x3=18
3. 1+3x3+3x3+3x3=28 cont.
4. 2+3x3+3x3=20
5. 2+3x3=11

Do this problem

 333
X333

1.
2.
3.
4.
5.

Practice: Study this problem

444
<u>444</u>

1. 4x4=16
2. 1+4x4+4x4=33
3. 3+4x4+4x4+4x4=51 cont.
4. 5+4x4+4x4=37
5. 3+4x4=19

Final Answer: 197,136

Do this Problem:

 444
X444

1.
2.
3.
4.
5.

<u>Review for Accuracy</u>

<u>Do 34 times tables 34x5=</u>   34x7=   34x9=   34x6=

34x1=34
34x2=68
34x3=102
34x4=136
34x5=170
34x6=204
34x7=238
34x8=272
34x9=306
34x10=340
34x11=374
34x12=408

Do 45 times tables

45x1=45
45x2=90
45x3=135
45x4=180
45x5=225
45x6=270
45x7=315
45x8=360
45x9=405
45x10=450
45x11=495
45x12=540

Do 39 time tables

Check time tables 39x12=_____39x7=_____39x5=_____39x11=_____
39x10=_____

39x1=39
39x2=78
39x3=117
39x4=156
39x5=195
39x6=234
39x7=273
39x8=312
39x9=351
39x10=390
39x11=429
39x12=468

Do timetables 44

Check 43 times tables Review for Accuracy 43x5=_____ 43x6=_____ 43x7=_____
43x9=_____

43x1=43
43x2=86
43x3=129
43x4=172
43x5=215
43x6=258
43x7=301
43x8=344
43x9=387
43x10=430
43x11=473
43x12=516

Check 44 time tables

44x1=44
44x2=88
44x3=132
44x4=176
44x5=220
44x6=264
44x7=308
44x8=352
44x9=396
44x10=440
44x11=484
44x12=528

PART XIII

PULLING IT ALL TOGETHER

The easiest Method to do Mental Multiplication:

1111   1-2-3-4-3-2-1

<u>1111</u>   1. 1x1=

          2. 1x1+1x1= 2

          3. 1x1+1x1+1x1=3

          4. 1x1+1x1+1x1+1x1=4

          3. 1x1+1x1+1x1=3

          2. 1x1+1x1=2   Cont.

          1. 1x1=1

          Answer: 1,234,321

Check: 1234321

1 1 1 1 1 1 1

<u>1 2 3 4 3 2 1</u> Multiply

1, 2 3 4 ,3 2 1 (No need to add to these numbers. These are the correct answer)

2222   1-2-3-4-3-2-1- Directions

<u>2222</u>   1. 2x2=4

          2. 2x2+2x2=8

          3. 2x2+2x2+2x2=12

          4. 1+2x2+2x2+2x2+2x2=17

          3. 1+ 2x2+2x2+2x2=13

          2. 1+ 2x2+2x2=9

          1. 2x2=4

          Answer: 4,937,284

<u>Cont.</u>

<u>Check</u>

<u>Find the Greatest Common Factor: "4"</u>

4 4 4 4 4   4   4 Multiply

<u>1 2 3 4 3   2   1</u>

4 8 12 16 12 8   4 Add

<u>   1 1   1</u>

<u>4</u> <u>8</u> 12 <u>17</u> 13 <u>9</u> 4 (backward)

4 8 12 16 12 8 4

<u> 1   1   1</u>

<u>4</u> <u>9</u> 13 <u>17</u> 12 <u>8</u> <u>4</u> Final Answer:4,937,284

3333   Directions: 1-2-3-4-3-2-1

<u>3333</u>   1. 3x3=9 cont.

         2. 3x3+3x3=18 Cont.

         3. 1+3x3+3x3+3x3=28

         4. 2+3x3+3x3+3x3+3x3=38

         3. 3+3x3+3x3+3x3=30

         2. 3+3x3+3x3=21

         1. 2+3x3=11

         Answer: 11,108,889 Cont.

Check 1-2-3-4-3-2-1 Greatest Common Factor: "9"

9 9   9   9   9   9 9 Multiply

<u>1 2 3   4   3   2 1</u>

9 18 27 36 27 18 9 Add

<u>   1   2   2   3   2</u>

<u>9</u> 18 <u>28</u> <u>38</u> <u>30</u> 21 <u>1</u> 1 (backward)

                           Final Answer:11,108,889

4444   1-2-3-4-3-2-1

<u>4444</u>   1. 4x4=16

         2. 1+ 4x4+4x4=33

         3. 3+4x4+4x4+4x4=51

         4. 5+4x4+4x4+4x4+4x4=69

         3. 6+4x4+4x4+4x4=54

         2. 5+4x4+4x4=37

         1. 3+4x4=19

         Answer: 19,749,136

Check

<u>1234321</u>

16 16 16   16 16   16 16 Greatest Common Factor

<u>1   2   3   4   3   2   1 Multiply</u>

16 32 48   64 48   32 16

<u>   1   3   5   6   5   3   Add</u>

16 <u>33</u> 51   <u>69</u> <u>54</u>   <u>37</u> <u>19</u> (backward) Final Answer: 19,749,136

5555   1-2-3-4-3-2-1
5555   1. 5x5=25
            2. 2+5x5+5x5=52
            3. 5+5x5+5x5+5x5=80
            4. 8+ 5x5+5x5+5x5+5x5=108
            3. 10+5x5+5x5+5x5=85
            2. 8+5x5+5x5=58
            1. 5+5x5=30
            Answer: 30,858,025

Check 1-2-3-4-3-2-1

25	25	25	25	25	25	25	
1	2	3	4	3	2	1	Multiply
25	50	75	100	75	50	25	
	2	5	8	10	8	5	Add
25	52	80	108	85	58	30	Answer: 30,858,025

Cont.

Reversal

25	50	75	100	75	50	25	
5	8	10	8	5	2		
30	58	85	108	80	52	25	Final Answer: 30,858,025

Try this problem

6666   1-2-3-4-3-2-1
6666   1. 6x6=36
44,435,556   2. 6x6+6 x 6+3=75
                 3. 7+6x6+6x6+6 x 6=115
                 4. 5+ 6x6+6x6+6x6+6 x 6=155
                 3. 14+6x6+6x6+6 x 6=123
                 2. 11+6x6+6x6=84
                 1. 8+6x6=44
                 Answer; 44,435,556

Check 1-2-3-4-3-2-1

36	36	36	36	36	36	36	
1	2	3	4	3	2	1	Multiply
36	72	108	144	108	72	36	Add
	3	7	11	15	12	8	
36	75	115	155	123	84	44	(backward)

                            Final Answer: 44,435,556

```
7777      1-2-3-4-3-2-1
7777      1.  7x7=49
          2.  4+7x7+7x7=102
          3.  10+7x7+7x7+7x7=157 Cont.
          4.  15+7x7+7x7+7x7+7x7=211
          5.  21+7x7+7x7+7x7=168
          6.  16+7x7+7x7=114
          Answer: 60,481,729
```

Check 1-2-3-4-3-2-1

```
49   49    49    49    49    49     49  Multiply
1    2     3     4     3     2      1
49   98    147   196   147   98     49  Add
     4     10    15    21    16     11
49   102   157   211   168   114    60
                       Answer:60,481,729
```

```
8888      1-2-3-4-3-2-1
8888      1.  8x8=64
          2.  6+8x8+8x8=134
          3.  13+8x8+8x8+8x8=205
          4.  20+8x8+8x8+8x8+8x8=276
          3.  27+8x8+8x8+8x8=219
          2.  21+8x8+8x8=149
          1.  14+8x8=78
          Answer: 78,996,544
```

Check 1-2-3-4-3-2-1

```
64    64    64    64    64    64    64
1     2     3     4     3     2     1 Multiply
64   128   192   256   192   128   64
      6    13    20    27    21    14 Add
64   134   205   276   219   149   78 cont.
                       Answer: 78,996,544
```

```
9999    1-2-3-4-3-2-1
9999    1.  9x9=81
        2.  8+9x9+9x9=170
        3.  17+8x8+8x8+8x8=260
        4.  26+8x8+8x8+8x8+8x8=350
        3.  35+8x8+8x8+8x8=278
        2.  27+8x8+8x8=189
        1.  18+81=99 cont.
        Answer: 99,980,001
```

Check 1-2-3-4-3-2-1

```
81  81   81   81   81  81  81
 1   2    3    4    3   2    1 Multiply
81  162 243  324  243 162 81 Add
      8  17   26   35   27 18
81  170 260  350  278 189  99 Answer:99,980,001
```

Note:

The above processes are not limited to four digits. This will work for many numbers of digits.
Try it!

```
11111   1-2-3-4-5-4-3-2-1 (Figure out the code!)
11111   1.  1x1=1
        2.  1x1+1x1=2
        3.  1x1+1x1+1x1=3
        4.  1x1+1x1+1x1+1x1=4
        5.  1x1+1x1+1x1+1x1+1x1=5
        4.  1x1+1x1+1x1+1x1=4
        3.  1x1+1x1+1x1=3
        2.  1x1+1x1=2
        1.  1x1=1
        Answer: 123,454,321 (Check will elicit the same answer)
```

22222   1-2-3-4-5-4-3-2-1
<u>22222</u>   1. 2x2=4
           2. 2x2+2x2=8
           3. 2x2+2x2+2x2=12
           4. 1+2x2+2x2+2x2+2x2=17
           5. 1+2x2+2x2+2x2+2x2+2x2=21
           4. 2+2x2+2x2+2x2+2x2=18
           3. 1+2x2+2x2+2x2=13
           2. 1+2x2+2x2=9
           1. 2x2=4
           Answer: 493,817,284

Check: 1-2-3-4-5-4-3-2-1

4	4	4	4	4	4	4	4	4
<u>1</u>	<u>2</u>	<u>3</u>	<u>4</u>	<u>5</u>	<u>4</u>	<u>3</u>	<u>2</u>	<u>1</u>
4	8	12	16	20	16	12	8	4
			<u>1</u>	<u>1</u>	<u>2</u>	<u>1</u>	<u>1</u>	
<u>4</u>	<u>8</u>	<u>12</u>	<u>17</u>	<u>21</u>	<u>18</u>	<u>13</u>	<u>9</u>	<u>4</u>

33333   1-2-3-4-5-4-3-2-1
<u>33333</u>   1. 3x3=9
           2. 3x3+3x3=18
           3. 1+3x3+3x3+3x3=28
           4. 2+3x3+3x3+3x3+3x3=38
           5. 3+3x3+3x3+3x3+3x3+3x3=48
           4. 4+3x3+3x3+3x3+3x3=40
           3. 4+3x3+3x3+3x3=31
           2. 3+3x3+3x3=21 cont.
           1. 2+3x3=11 cont.
           Answer: 1,111,088,889

Check 1-2-3-4-5-4-3-2-1

Cont.

9	9	9	9	9	9	9	9	9
<u>1</u>	<u>2</u>	<u>3</u>	<u>4</u>	<u>5</u>	<u>4</u>	<u>3</u>	<u>2</u>	<u>1</u>
9	18	27	36	45	36	27	18	9
		<u>1</u>	<u>2</u>	<u>3</u>	<u>4</u>	<u>4</u>	<u>3</u>	<u>2</u>
<u>9</u>	<u>18</u>	<u>28</u>	<u>38</u>	<u>48</u>	<u>40</u>	<u>31</u>	<u>21</u>	<u>11</u>

Answer: 1,111,088,889

4,444X4,444=1,975,269,136

1. 4x4=1<u>6</u>
2. 1+4x4+4x4=3<u>3</u>
3. 3+4x4+4x4+4x4=5<u>1</u>
4. 5+4x4+4x4+4x4+4x4=6<u>9</u>
5. 6+4x4+4x4+4x4+4x4+4x4=8<u>6</u>
4. 8+ 4x4+4x4+4x4+4x4=7<u>2</u>
3. 7+4x4+4x4+4x4=5<u>5</u>   Cont.
2. 5+4x4+4x4=3<u>7</u>
1. 3+4x4=<u>19</u>

Answer: 1,975,269,136

Check: <u>16   16   16   16   16   16   16   16   16 Greatest Common Factor</u>
         1   2   3   4   5   4   3   2   1 Multiply
        16 32   48   64   80   64   48 32   16
         <u>1   3   5   6   8   7   5   3 Add</u>
        16 33   51   69   86   72   55   37 19 Answer backward 6,319,625,791

Answer Forward: 1,975,269,136

Reversal

16   32   48 64   80   64   48 32   16
<u>3   5   7   8   6   5   3   1</u>
<u>19 37   55   72 86   69   51   33 16 Final Answer:1,975,269,136</u>

<u>Do timetables 51</u>

51x1=51 51x3=_____51x6=_____51x7=_____51x8=_____
51x2=102
51x3=153
51x4=204
51x5=255
51x6=306
51x7=357
51x8=408
51x9=459
51x10=510
51x11=561
51x12=612

Do times tables 41

Check 41 timetables 41x6=_____ 41x7=_____ 41x8=_____ 41x10=_____

41x1=41
41x2=82
41x3=123
41x4=164
41x5=205 cont.
41x6=246
41x7=287
41x8=328
41x9=369
41x10=410
41x11=451
41x12=492

```
   555
   555
   555
_____
308,025
```

1. 5x5=2<u>5</u>
2. 2+5x5 =5x5=5<u>2</u>
3. 5+5x5+5x5+5x5=8<u>0</u>
4. 8+25+25=5<u>8</u>
5. 5+25=<u>30</u>
Final Answer: 308,025

Mystery Multiplication

```
25  25  25  25  25
 1   2   3   2   1
25  50  75  50  25
 1   2   5   8   5
25  52  80  58  30  Answer backward= 520,830
```

Reversal

```
25   50  75  50 25
 5    8   5   2  1
30  58  80  52  25  Final Answer: 308,025 forward
```

Check 42 timetables 42x2=_____ 42x5=_____ 42x7=_____ 42x 8=_____

42x1=42
42x2=84
42x3=126
42x4=168
42x5=210
42x6=252
42x7=294
42x8=336
42x9=378
42x10=420
42x11=462
42x12=504

Extra Examples:

No.1

6666
<u>6666</u>
Answer: 44,435,556

1. 6x6=3<u>6</u>
2. 6x6+6x6=72+3=7<u>5</u>
3. 6x6+6x6+6 x 6=108+ 7=11<u>5</u>
4. 6x6+6x6+6x6+6 x 6=144+11=15<u>5</u>
5. 6x6+6x6+6 x 6=108+15=12<u>3</u>
6. 6x6+6x6=72+12=8<u>4</u>
7. 6x6=36+8=<u>44</u>
<u>Answer: 44,435,556</u>

<u>Continued</u>

<u>N0. 2</u>

<u>36</u>	<u>36</u>	<u>36</u>	<u>36</u>	<u>36</u>	<u>36</u>	<u>36 Greatest Common Factor</u>
<u>1</u>	<u>2</u>	<u>3</u>	<u>4</u>	<u>3</u>	<u>2</u>	<u>1 Multiply</u>
36	72	108	144	108	72	36
	<u>3</u>	<u>7</u>	<u>11</u>	<u>15</u>	<u>12</u>	<u>8 Add</u>
3<u>6</u>	7<u>5</u>	11<u>5</u>	15<u>5</u>	12<u>3</u>	<u>84</u>	<u>44</u> Answer: 44,435,556

7777
<u>7777</u>
60,481,729

1. 7x7=4<u>9</u> cont.
2. 7x7+7x7=98+4=10<u>2</u>
3. 7x7+7x7+7x7=147+10=15<u>7</u>
4. 7x7+7x7+7x7+7x7=196+15=21<u>1</u>
5. 7x7+7x7+7x7=147+21=16<u>8</u>
6. 7x7+7x7=98+16=11<u>4</u>
7. 7x7=49+11=<u>60</u>
Answer: 60,481,729

49	49	49	49	49	49	49 Greatest Common Factor
<u>1</u>	<u>2</u>	<u>3</u>	<u>4</u>	<u>3</u>	<u>2</u>	<u>1 Multiply</u>
49	98	147	196	147	98	49
	<u>4</u>	<u>10</u>	<u>15</u>	<u>21</u>	<u>16</u>	<u>11 Add</u>
4<u>9</u>	10<u>2</u>	15<u>7</u>	21<u>1</u>	16<u>8</u>	11<u>4</u>	60

Answer: 60,481,729

Reversal

49	98	147	196	147	98	49
<u>11</u>	<u>16</u>	<u>21</u>	<u>15</u>	<u>10</u>	<u>4</u>	
<u>60</u>	10<u>4</u>	16<u>8</u>	20<u>1</u>	15<u>7</u>	10<u>2</u>	4<u>9</u> Final Answer: 60,481,729

8,888
<u>8,888</u>
78,996,544

1. 8x8=64
2. 8x8+8x8=128+6=134
3. 8x8+8x8+8x8=192+13=205
4. 8x8+8x8+8x8+8x8=256+20 =276
5. 8x8+8x8x8x8=192+27=219
6. 8x8+8x8=128+21=149
7. 8x8=64+14=78
Answer: 78,996,544

Cont.

64	64	64	64	64	64	64 Greatest Common Factor
1	2	3	4	3	2	1 Multiply
64	128	192	256	192	128	64
	6	13	20	27	21	14 Add
64	134	205	276	219	149	78

Cont.

Reversal

64 128 192 256 192 128 64
14 21 27 20 13   6
78 149 219 276 205 134 64

9999
9999

1. 9x9=81
2. 9x9+9x9=162+8=170
3. 9x9+9x9+9x9=+17=260
4. 9x9+9x9+9x9+9x9+26=350 Cont.
5. 9x9+9x9+9x9+35=278
6. 9x9+9x9+27=189
7. 9x9=81+18=99 Cont.
Answer: 99,980,001

81	81	81	81	81	81	81 Greatest Common Factor
1	2	3	4	3	2	1 Multiply
81	162	243	324	243	162	81
	8	17	26	35	27	18 Add
81	170	260	350	278	189	99 (backward)

Answer:99,980,001 cont.

Reversal

81 162 243 324 243 162 81
18 27 35 26 17   8
99 189 278 350 260 170 81 Final Answer: 99,980,001

The hardest problems in this book are as follows:

```
  222222
x   2222
493,777,284
```

1. 2x2=4
2. 2x2+2x2=8
3. 2x2+2x2+2x2=12
4. 1+2x2+2x2+2x2+2x2=17
5. 1+2x2+2x2+2x2+2x2=17
6. 1+2x2+2x2+2x2+2x2=17 cont.
7. 1+2x2+2x2+2x2=13
8. 1+2x2+2x2=9
9. 2x2=4

```
  333333
x  22222
740,665,926
```

1. 2x3=6
2. 3x2+2x3=12 cont.
3. 1+2x3+2x3+2x3=9
4. 1+2x3+2x3+2x3+2x3=25
5. 2+2x3+2x3+2x3+2x3=26
6. 2+2x3+2x3+2x3+2x3=26
7. 8x6+2x3=20
8. 3x2+2+2x3=14
9. 1+2 x 3=7
Answer: 740,665,926

```
444444
x 2222
987,554,568
```

1. 2x4=8
2. 2x4+2x4=16 cont.
3. 1+2x4+2x4+2x4=25
4. 2+2x4+2x4+2x4+2x4=34
5. 2+2x4+2x4+2x4+2x4=35
6. 3+2x4+2x4+2x4+2x4=35 Cont.
7. 3+2x4+2x4+2x4=27
8. 2+2x4+2x4=18
9. 1+2x4=9
Answer: 987,554,568

101,010
x 2222
224,444,220

1. 2x0=0
2. 2x0+2 x 1=2
3. 2x0+2x0+2 x 1=2
4. 2 x 0+2x1+2x0+2 x 1=4
5. 2 x 0+2x1+2x0+2 x 1=4
6. 2 x 0+2x1+2x0+2 x 1=4
7. 2x1+2x1=4
8. 2x1+2x0=2 cont.
9. 2x1+2x0=2
Answer: 224,444,220

More difficult problems:

11111111
    1111
12,344,444,321

Procedure:
1. Multiply the first four1's: you will receive 4,321.
2. Move to the 5th "1" 6th, "1" 7th "1" 8th "1" add the remaining "1's and you will receive this number: 12,344,444,321.

12121212
    1111
13,466,666,532

Procedure:
1. Multiply the first four "1's" you will receive 6,532.
2. Move to the 5th "1" 6th etc when you run out of 1's then add the remaining "1's' and 2's, you receive this number: 13,466,666,532

13131313
    1111
14,588,888,743

Procedure:
1. Multiply the first four 1's and you will receive the numbers 8,743.
2. Move to the 5th "1", 6th etc. when you run out of 1's add the remaining 1's and 3's, you will receive this number 14,588,888,743

14141414
___1111___
15,711,110,954

Procedure:
1. Multiply the first four 1's and you will receive the numbers, 0,954.
2. Move to the 5th 1', 6th 1' etc, when you run out of 1's, add the remaining 1's and 4's and you will receive the numbers 15,711,110,954.

15151515
___1111___
16,833,333,165

Procedure
1. Multiply the four 1's and you will receive the numbers, 3,165.
2. When you run out of 1's you will have this number 3,165.
   When you run out of 1's add the remaining 1's and 5;s, and you will receive this number: 16,833,333,165.

16161616
___1111___
17,955,555,373

Procedure
1. Multiply the four 1's and you will receive the numbers 5,373
2. When you run out of 1's add the remaining 6's and 1's and you will receive this number: 17,955,555,373.

Another Mystery Multiplication

```
      1111                1,234,321   (use simple multiplication to solve problem)
      1111               11,111,111
      1111                1,234,321  (Key)  Achieves the same answer
      1111
      1111
      1111
 1,234,321   (Key)
```

```
      1111                1,234,321   (Key)(simple multiplication will give the answer to the problem
      2222              x 2,222,222     on the left)
      2222                2,468,642
      2222
      2222
      2222
 2,468,642
```

```
      1111                1,234,321   (Key)  Just simple multiplication will answer the problem on the left)
      3333                3,333,333
      3333                3,702,963
      3333
      3333
      3333
 3,702,963
```

```
      1111                1,234,321  (Key) use simple multiplication to obtain the answer the problem
      4444                4,444,444     to the left)
      4444                4,937,284
      4444
      4444
      4444
 4,937,284
```

```
    1111              1,234,321 (use simple multiplication to solve the problem on the left)
    5555              5,555,555
    5555              6,171,605
    5555
   5555
  5555
6,171,605


    1111
    6666              1,234,321 (Key)
    6666              6,666,666
   6666              7,405,926
  6666
 6666
7,405,926


    1111              1,234,321
    7777              7,777,777
   7777              8,640,247
  7777
 7777
7777
8,640,247
```

My Beautiful Golden Nuggets: Step 1. Multiply "4" x 1,234,321= 4,937,284

1)   2222 (coding, 2x2,2x2,2x2,2x2=4444)   1234321   My way: Simple Multiplication. Step 2
     x 2222   4   will give you the same answer
       4444   4,937,284
       4444
       4444
       4444
4, 9 3 7, 2 8 4

Step 1.Multiply " 9" x 1,234,321= 11,108,889

3333 (coding, 3x3, 3x3, 3x3, 3x3)   1,234,321 My way: Simple Multiplication Step 2 gives you the
3333   9   same answer
    9999   11,108,889
    9999
9999
9999
11,108,889

Step 1	Step 2	Step 3	
3333	1,234,321= Key	3333	
2222	x 6	2222	
6666	7, 4 0 5, 9 2 6	7,405,926	1)2x3=6
6666			2)2x3+2x3=12
6666			3)1+2x3+2x3+2x3=19
6666			4)1+ 2x3+2x3+2x3+2x3=25
7,405,926			5)2+2x3+2x3=14
			6)2x3+1=7
			Final Answer: 7,405,926

PART XIV
UNIQUE PROBLEMS

Unique problems:
Use this "Key" 1,234,321

```
        1111
        2222
        2222
        2222
        2222
        2222
2,468,642 (double the key)
```

"Key"
```
1,234,321    ( simpler way)
        2
2,468,642
```

```
1111
3333
    3333
   3333
  3333
3333
3,702,963
```

Use "Key"
```
1,234,321
        3
3,702,963
```

```
3333
2222
    6666
   6666
  6666
6666
7,405,926

1,234,321 (key)
          6
7,405,926
```

```
4444
2222
    8,888
   8,888
  8,888
 8,888
9,874,568

1,234,321 (key)
          8
9,874,568
```

```
2222
2222
    4444
   4444
  4444
 4444
4,937,284
```

```
"Key"
1,234,321
          4
4,937,284
```

Cont.

1,111
<u>5,555</u>
    5555
   5555
  5555
<u>5555</u>
6,171,605
"Key"
1,234,321
<u>   5</u>
6,171,605

1,111
<u>6,666</u>
    6666
   6666
  6666
<u>6666</u>
7,405,926
"Key"
1,234,321
<u>   6</u>
7,405,926

   3,333
<u> 2,222</u>
  6,666
  6,666
  6,666
<u>6,666</u>
7,405,926

1,234,321
<u>   6</u>
7,405,926

2,468,642 (double the key)

_____3

7,405,926

   2222

_____2222

     4444

    4444

   4444

  4444

4,937,284

2,468,642 (double the key)

_____2

4,937,284

PART XV

MENTAL DIVISION

2/2=1
2/22=11
2/222=111
2/2222=1,111
2/22222=11,111
2/222222=111,111
2/2222222=1,111,111
2/22222222=111,111,111
2/222222222=111,111,111,111
2/2222222222=1,111,111,111

4/4=1
4/44=11
4/444=111
4/4,444=1,111
Acceleration
5/5=1
5/55=11
5/555=111

Cont.
5/50=10
50/500=10
500/5,000=10
<u>5000/50,000=10</u>
6/6=1
60/60=1
600/600=1
Acceleration
60/600=10
600/6,000=10
6,000/6,000=1
Acceleration:
10/100=10
100/1000=10
<u>1,000/10,000=10</u>
15/30=2
150/300=2
1,500/3,000=2
20/200=10
200/2,000=10
2,000/20,000=10

25/50=2
250/500=2
2,500/5,000=2

Please study these facts

Division Made Simple: Steps to Mental Division

2/2=1
2/22=11
2/222=111
2/2,222=1,111
2/22,222=11,111
2/222,222=111,111
3/3=1
3/33=11
3/333=111
3/3,333=1,111
Start here:
4/4=1
4/44=11
4/444=111
4/4,444=1,111
4/8=2
4/16=4 cont.
4/32=8
4/40=10
4/80=20
5/5=1
5/10=2
5/15=3
5/20=4
5/50=10
5/100=20
6/6=1
6/12=2
6/30=5
6/36=6
6/42=7
6/48=8 cont.
6/54=9
6/60=10
6/66=11
6/72=12
7/56=8

7/63=9
7/70=10
8/80=10
8/88=11
9/108=12
20/40=2
3/30=10
4/40=10
5/50=10
6/60=10
7/70=10
8/80=10
9/90=10
10/100=10
11/11=1

Greatest Common Factor

The multiplicand must be multiplied by the multiplier. Hence the multiplier when multiplied by the multiplicand will yield the product of division, the Quotient. The result will be the correct answer to the problem consistently i.e. Divisor-2, Dividend-10, Quotient 5, 2/10=5 hence(multiplier) 2 x multiplicand 5 = 10 the Quotient.
Again
Multiplier 11, divisor when divided into 121 will yield the Quotient 11.
11/121 (11x11=121) 11, the Multiplier 11x the Multiplicand 11 will yield the product 121.
Study this.

11/121=11 Cont.
12/24=2 (2x12=24)
12/48=4(4x12=48)
13/26=2(2x12=26)
13/52=4 13/52=?Put answer here
14/28=2 14/28=? Put answer here
14/56=4 14/56= ?Put answer here
15/30=2 2x15=?
15/60=4 4x15=?
16/32=2 2x16=?
16/64=4Study
17/34=2 Study
17/78=4Study
18/36=2 Study
18/72=4Study
19/38=2 Study Cont.
19/76=4Study

Harder problems

2/444=222
3/666=222
4/121212=30,303
Easy to hard
2/2=1
2/4=2
2/6=3
2/8=4
2/10=5
2/12=6
2/14=14
2/16=8
Acceleration
3/21=7 cont.
3/2,121=707
3/212,121=70,707
4/24=6
4/2,424=606
4/242,424=60,606
Acceleration
5/20=4
5/2,020=404
5/202,020=40,404
6/36=6
6/3,636=606
6/363,636=60,606
7/14=2
7/1,414=202
7/1,414,14=20,202
8/16=2
8/1,616=202
8/161,616=20,202
9/181,818=20,202
10/20=2
10/2,020=202
10/202,020=20,202

Please study these mental division problems. The key to many long division problems are the Greatest Common Factors found in the problems.

2222
<u>9999</u>
22,217,778

Greatest Common Factor: 2/9=18

```
18  18  18  18  18  18  18
1    2   3   4   3   2   1
18  36  54  72  54  36  18
     1   3   5   7   6   4
18  37  57  77  61  42  22   Answer:  22,217,778  Right to Left
```

Reversal

```
18    36    54    72    54    36  18
 4     6     7     5     3     1
22    32    51    77    67    47  18   Answer: 22,217,778 Left to Right
```

CONCLUSION

I thoroughly *enjoyed* writing this book for all you readers who want to make the world a little better with a little help from your friend that sticks closer than a brother. His name is Jesus. I caution you to use the incredible mechanism within you called "the brain." The mechanical mechanism called the calculator should not take the place of what each one of you are endowed with that can be used to change the environment in which you live and ultimately the world in which you engage.

Mental Multiplication Vol 2 comes to an end now. This book was very insightful to write. It was exciting and engaging. It is my desire that multiplication without the use of calculators or computers except to confirm your answer will be the new math for the students of mathematics.

Thank you, and God bless you, in Jesus' perfect and blessed Name.

www.ingramcontent.com/pod-product-compliance
Lightning Source LLC
Chambersburg PA
CBHW081437170526
45166CB00008B/2227